Jolanta Klocek

Thin carbon containing films

Jolanta Klocek

Thin carbon containing films

Processing, investigation and possible application

Südwestdeutscher Verlag für Hochschulschriften

Impressum/Imprint (nur für Deutschland/only for Germany)
Bibliografische Information der Deutschen Nationalbibliothek: Die Deutsche Nationalbibliothek verzeichnet diese Publikation in der Deutschen Nationalbibliografie; detaillierte bibliografische Daten sind im Internet über http://dnb.d-nb.de abrufbar.
Alle in diesem Buch genannten Marken und Produktnamen unterliegen warenzeichen-, marken- oder patentrechtlichem Schutz bzw. sind Warenzeichen oder eingetragene Warenzeichen der jeweiligen Inhaber. Die Wiedergabe von Marken, Produktnamen, Gebrauchsnamen, Handelsnamen, Warenbezeichnungen u.s.w. in diesem Werk berechtigt auch ohne besondere Kennzeichnung nicht zu der Annahme, dass solche Namen im Sinne der Warenzeichen- und Markenschutzgesetzgebung als frei zu betrachten wären und daher von jedermann benutzt werden dürften.

Coverbild: www.ingimage.com

Verlag: Südwestdeutscher Verlag für Hochschulschriften GmbH & Co. KG
Heinrich-Böcking-Str. 6-8, 66121 Saarbrücken, Deutschland
Telefon +49 681 37 20 271-1, Telefax +49 681 37 20 271-0
Email: info@svh-verlag.de

Approved by: Cottbus, TU, Diss. , 2012

Herstellung in Deutschland (siehe letzte Seite)
ISBN: 978-3-8381-3266-2

Imprint (only for USA, GB)
Bibliographic information published by the Deutsche Nationalbibliothek: The Deutsche Nationalbibliothek lists this publication in the Deutsche Nationalbibliografie; detailed bibliographic data are available in the Internet at http://dnb.d-nb.de.
Any brand names and product names mentioned in this book are subject to trademark, brand or patent protection and are trademarks or registered trademarks of their respective holders. The use of brand names, product names, common names, trade names, product descriptions etc. even without a particular marking in this works is in no way to be construed to mean that such names may be regarded as unrestricted in respect of trademark and brand protection legislation and could thus be used by anyone.

Cover image: www.ingimage.com

Publisher: Südwestdeutscher Verlag für Hochschulschriften GmbH & Co. KG
Heinrich-Böcking-Str. 6-8, 66121 Saarbrücken, Germany
Phone +49 681 37 20 271-1, Fax +49 681 37 20 271-0
Email: info@svh-verlag.de

Printed in the U.S.A.
Printed in the U.K. by (see last page)
ISBN: 978-3-8381-3266-2

Copyright © 2012 by the author and Südwestdeutscher Verlag für Hochschulschriften GmbH & Co. KG and licensors
All rights reserved. Saarbrücken 2012

„ Learn from yesterday, live for today, hope for tomorrow.

The important thing is not to stop questioning."

„I never think of the future. It comes soon enough."

Albert Einstein

Kurzfassung

In dieser Arbeit werden detaillierte Untersuchungen zur Herstellung und Stabilität von dünnen kohlenstoffhaltigen Filmen und ihre mögliche Anwendung als Materialien mit niedriger Dielektrizitätskonstante (low k) präsentiert. Um komplexe Informationen über die chemischen, morphologischen und dielektrischen Eigenschaften der erzeugten Schichten zu erhalten, wurde eine Kombination verschiedener Methoden angewendet. Dabei kamen spektroskopische (Röntgen-Photoelektronen-Spektroskopie: XPS, Röntgen-Nahkanten-Absorptions-Spektroskopie: NEXAFS und Fourier-Transform-Infrarot-Spektroskopie: FTIR), mikroskopische (Rasterkraftmikroskopie: AFM) und elektrische (Kapazitäts-Spannungs-Messungen: CV) Charakterisierungsmethoden zum Einsatz.

Die Filme wurden mit verschiedenen Techniken aufgetragen: Verdampfung, Sprüh- und Tropfverfahren sowie Spin Coating. Bezüglich der möglichen low-k-Anwendung wurden hybride organisch-anorganische kohlenstoffhaltige Verbundwerkstoffe untersucht. Die Filme basieren auf 3-Aminopropyltrimethoxysilan (APTMS), das mit den folgenden Dotierungen angereichert wurde: C_{60} Fulleren, Phenyl-C_{61}-Buttersäure-methylester (PCBM), Kupfer-Phthalocyanin (CuPc) und Tris (dimethylvinylsilyloxy)-POSS (POSS).

Die vorliegende Dissertation stellt progressive Optimierungsschritte dar, mit denen eine allmähliche Verringerung der resultierenden Dielektrizitätskonstante des Hybrid-Materials erreicht wurde. Wie die durchgeführten Untersuchungen zeigen, erlaubt die Substitution von C_{60} in der APTMS Matrix durch sein besser lösliches Derivat PCBM die Erhöhung der Konzentration der Kohlenstoffspezies innerhalb der Schichten. Die Einführung von POSS als zusätzliche Dotierung führte zu einer Erhöhung der Resistenz des produzierten Materials gegen Umgebungseinflüsse. Jedoch wurden dabei ab bestimmten POSS-Konzentrationen (im Bereich von 0.3 %) Überladungen der Matrix mit dem Dotierstoff festgestellt, was sich in der Bildung fraktaler Strukturen widerspiegelte. Letztendlich erzielte die Kombination relativ niedriger Konzentration von CuPc und POSS-Molekülen (im Bereich von 0.1 %), die innerhalb der APTMS Matrix dispergiert wurden, die besten Ergebnisse und es konnten homogene Schichten mit einer ultra-niedrigen Dielektrizitätskonstante von 1.8 hergestellt werden.

Abstract

In this thesis the detailed investigations concerning processing and stability of thin films including carbon species and their possible application as materials of a low dielectric constant (low k) are presented. In order to gather a complex information regarding the chemical, morphological and dielectric properties of the produced layer a combination of the spectroscopy: X-ray photoelectron spectroscopy (XPS), near edge X-ray absorption fine structure spectroscopy (NEXAFS) and Fourier transform infrared spectroscopy (FTIR), microscopy: atomic force microscopy (AFM) and electrical characterization: capacitance-voltage technique (CV) have been applied.

The films deposited by means of variety of techniques have been described, ranging from evaporation, through spraying and dropping to spin-coating. Regarding the possible low-k application, a considerable attention has been paid to the hybrid organic-inorganic 3-aminopropyl-trimethoxysilane (APTMS) based composite materials enriched with carbon species coming from the following dopants: C_{60} fullerenes, [6,6]-phenyl-C_{61}-butyric acid (PCBM), copper phthalocyanine (CuPc), and tris(dimethylvinylsilyloxy)-POSS (POSS).

In the following thesis progressive steps leading to gradual decreasing of the resulting permittivity of the hybrid material is presented. As revealed by the performed investigations, the replacement of C_{60} within the APTMS based matrix by its better soluble derivative PCBM allows the increase of the concentration of the carbon species within the composite films. The introduction of POSS as an additional dopant gave the opportunity of increasing the resistance of the produced material against the ambient influence. With the excess of the POSS concentration an original fractal-shaped cluster formation has been observed. Finally, the dispersion of the properly chosen low concentration of CuPc and POSS molecules within the APTMS based matrix led to the fabrication of homogenous layer with an ultra-low dielectric constant of 1.8.

Content

I. Introduction 1

II. Experimental details 6

    II.1. X-ray photoelectron spectroscopy 6

    II.2. NEXAFS spectroscopy 9

    II.3. FTIR spectroscopy 11

    II.4. AFM 12

    II.5. Capacitance-voltage measurement 13

    II.6. Deposition techniques 14

III. Results and discussion 18

    III.1. The stability of C_{60} and its derivatives upon handling in microsystems technologies 18

    III.2. Toluene influence on the thin fullerene and copper phthalocyanine film properties 22

    III.3. The stability of fullerol water solution in ambient condition 33

    III.4. Temperature influence on the properties of thin CuPc and C_{60} fullerene films deposited on silicon substrates 39

    III.5. Spectroscopic and atomic force microscopy investigations of the hybrid materials composed of the fullerenes and 3-aminopropyl-trimethoxysilane 46

    III.6. Fullerene based materials for ultra-low-k application obtained by the means of the sol-gel method 59

    III.7. Investigations of the chemical and electrical properties of fullerene and 3-aminopropyltrimethoxysilane based low-k materials 66

    III.8. Influence of the fullerene derivatives and cage polyhedral oligomeric silsesqiuoxanes on 3-aminopropyltrimethoxysilane based hybrid nanocomposites chemical, morphological and electrical properties 79

    III.9. Spectroscopic and capacitance-voltage characterization of thin aminopropyl-methoxysilane films doped with copper phthalocyanine, tris(dimethylvinylsilyloxy)-POSS and fullerene cages 102

IV. Conclusions and outlook 119

References 122

Appendix 133

    Frequently used abbreviations and symbols 133

Acknowledgements 134

I. Introduction

The great structural variety of the pure solid carbon forms including fullerenes, nanotubes, and graphite attract rapidly growing attention motivated by its potential importance in production of carbonaceous materials that could be applied in the field of catalyst supports, adsorbents, gas storage, electrode, carbon fuel cells, drug delivery and many others [1]. For a long time carbon has been known to exist in three forms: amorphous carbon (a), soft, black, stable graphite (b) and hard, transparent diamond (c). The revolutionary discovery of the C_{60} buckminsterfullerene in 1985, consisting of 12 pentagons and 20 hexagons and having a soccer-ball like structure, and subsequently carbon nanotubes in 1991 opened up entirely new horizons in materials science and nanotechnology due to their very interesting electronic, photonic, magnetic and mechanical properties. The carbon nanotechnology became an interdisciplinary field of science, including chemistry, physics, materials science, engineering and biology [2] and studies concerning its expanding developed with a tremendous rate.

Inert, hollow and indefinitely modifiable fullerenes seem to be promising candidates for many various potential applications, including superconductors, hydrogen storage, high-efficiency solar cells or chemical sensors [2]. Because of their high mechanical strength and low density fullerenes could find an application as the reinforcement of the super-strong and light-weight composite structures, formed from two or more distinct materials what provides desirable combinations of properties that are not found in the individual components [3]. Due to their unique properties, related to the integration of the organic and inorganic properties of the components, organic-inorganic hybrid materials are an object of intensive academic and industrial investigations [4]. Recently, the design and engineering of carbon-based materials attract increasing attention in the construction, aerospace and automotive industries. Carbon composites possess several advantages over traditional materials, for instance they do not undergo a corrosion process, are significant lighter than steel, and can be installed without the use of heavy construction equipment [5].

The interesting properties of the composite materials highly enriched with carbon species offer possibilities of its application also in microelectronics, where complex layered structures are deposited on a semiconductor surface to transmit intelligent electronic signals, forming an integrated circuit [6]. The microelectronics industry maintains the tendency of improving advanced processes productivity. This phenomenon follows the Moore's law predicting that the number of transistors in the integrated circuits increases twice every eighteen months [7]. However, because of the huge transistor density in the advanced integrated circuits its size has to be reduced what makes the circuits faster while at the same time the overall resistance-capacitance (RC) delay increases. One of the strategies to reduce the RC delay is the introduction of low-k materials as an isolation between the nano-wires in microelectronic devices [8,9]. Therefore, among a variety of materials building microelectronic structures are insulators, introduced as inter-metallic low-k dielectrics into the interconnect architecture, what allows the reduction of the parasitic capacitance, the power consumption and cross-talk of the microelectronic devices [10,11]. In the present thesis the deposition and characterization of some low-k hybrid composites containing fullerene cages and other carbon species, like copper phthalocyanine and POSS molecules functionalized with organic groups are presented.

The particular chapters of this thesis correspond to journal articles worked out during my research stage at the Chair of Applied Physics and Sensors of BTU Cottbus. Some of them are already published, others are just being reviewed, whereas the remaining parts will be submitted soon. The detail information concerning the submitted or already existing publications are given at the first page of the respective chapter.

The general information concerning the applied characterization techniques and the most common deposition methods used in this work is included in the chapter II.

The most crucial step in order to begin the experiments related to this work was the suitable samples preparation. Since the fullerenes, besides their obviously unique properties, seem to offer a wide processing flexibility, they were used as a component of the majority of the films described in this thesis. Moreover, a very important motivation of using fullerenes in this experiment is the fact that they are easy to detect and characterize by the means of microscopic and spectroscopic methods as well [12]. However, some problems related to the low C_{60} solubility had to be overcome as well as the proper deposition technique of the layers suitable for the further measurements had to be developed. Chapter III.1. contains the condensed information concerning the applied preparation techniques supported by the general spectroscopic results. This short overview contains the first and fundamental hints for the decision regarding the direction of investigations described in detail in the other chapters.

Chapter III.2. is a more detailed continuation of the chapter III.1. As it was already mentioned above, C_{60} fullerenes exhibit low solubility. Toluene is known to be commonly applied during the C_{60} processing, however the spectroscopic results described in chapter III.1. suggest that the fullerene molecules upon treatment with this solvent undergo some by-reactions. This process has been carefully investigated and interpreted in this part of the thesis. For comparison, also films composed of another carbon-reach molecule, copper phthalocyanine, upon toluene treatment were investigated. CuPc has been further applied as a component of the ultra-low *k* films described in the chapter III.9.

Since there are several processing problems related to the extreme hydrofobicity of C_{60}, much attention has been paid to its polihydroxylated and excellently water soluble derivative fullerol. Chapter III.3. describes a kind of an interesting curiosity concerning the high stability of this, unfortunately, not yet widely available material.

The chemical bonding between the carbon atoms within the C_{60} molecule is very strong, so their stability is high. However C_{60} moieties serve as excellent acceptors of the electrons what make them a great electrophilic reagent [12]. Primary and secondary aliphatic amines, due to their high nucleophilicity, easily undergo nucleophilic additions with C_{60} [13]. For that reason amine-functionalized siloxanes seem to be a proper material for the C_{60} functionalization, allowing the dispersion of the fullerene cages within the siloxane matrix. The fullerene functionalization with APTMS has already been reported in the literature. De Quan et al. obtained $C_{60}[APTMS]_n$ by dissolving C_{60} in APTMS at room temperature for 2 days [14]. Bell et al. produced similar adduct while stirring the APTMS mixture of C_{60}/C_{70} at 50 °C for 20 h [15]. Therefore the composition of the majority of the carbonaceous samples described in the following chapters consist of a high degree of this substance.

Many deposition techniques (for instance ALD) require the proper temperature treatment. Chapter III.4. reports on the heat treatment influence on the spin-coated composite material composed of the carbon dopants like fullerene cages and copper phthalocyanine dispersed in an APTMS based matrix. Aminoalkoxysilanes are used in such technologically important preparations techniques like ALD in order to functionalize the surface with amino groups and besides APTMS turned out to be an excellent medium allowing to obtain carbon reach spin-coated composite layers through simply, non-surfactant preparation route without external solvent application.

The chapter III.5. contains the characterization of the hybrid organic-inorganic composites obtained by the interesting combination of two deposition techniques: spin-coating and evaporation. As it was mentioned above, APTMS serves as an excellent source of the amino groups revealing at the same time great adhesion to the substrate. In this part of the thesis the deposition and measurements of the fullerene layer evaporated on the APTMS-coated substrate are reported. At this point one also find the first in this thesis relevant results concerning the application of the atomic force microscopy in the hybrid layers investigation. AFM turned out to be a perfect supplementary method to the spectroscopy, allowing the detailed investigation of the fullerenes molecules deposition onto the APTMS layer in applied experimental conditions. Interesting time-related migration of the amino groups originating from APTMS through the thick C_{60} evaporated layer is also reported.

After the development of the deposition techniques, the work focused more on the practical application of the produced films. Chapter III.6. has been written as a result of a successful cooperation with the group from the Wrocław University of Technology and opens a part of the thesis that is dedicated to the carbon-containing low dielectric constant materials. This part includes information regarding fullerene containing film of the low-k obtained by means of the sol-gel method. The sample has been produced by Dr. Justyna Krzak-Roś and Katarzyna Broczkowska from Wrocław University of Technology and also partially investigated by the Wrocław group. The k value of this sample achieved the promising value of 2.3 however its homogeneity was relatively low. Thus the improvement of the layer quality while omitting the sol-gel method became a starting point of the further researches described in the present thesis.

Beginning with the chapter III.7. successive steps of lowering the resulting permittivity of the hybrid APTMS based material enriched with the carbon species has been shown. Here exceptional abilities and professionalism of my collaborators (Dr. Karsten Henkel in the field of capacitance-voltage measurements (CV) and Dr. Krzysztof Kolanek in the field of AFM) opened a door for the gathering of complementary information concerning the properties of the produced low-k composite materials described in this thesis. The results reported in this chapter exhibit that the doping of the APTMS matrix with fullerene cages decreases its dielectric constant and the replacement of the C_{60} by its derivative: [6,6]-phenyl-C_{61}-butyric acid methyl ester (PCBM) leads to the increasing of the carbon species concentration and further lowering of the k value.

Chapter III.8. reports on further improvements of the low dielectric constant films. In this part of the researches POSS molecules possessing interesting hybrid inorganic-organic, three-dimensional structure have been applied as an additional dopant. The introduction of this component not only allowed to decrease the permittivity of the composite films but also increased their resistivity against the ambient influence. Apart from the low-k topic, for some described in this part samples the overloading of the POSS molecules within the compositions favored the heterogeneous phase separation and formation of the fractal shaped clusters giving the opportunity to obtain unusual AFM pictures, also presented in this chapter.

The final chapter III.9. is the culmination of the attempt to produce ultra-low-k APTMS based material enriched with the carbon species. As revealed by the investigation described in this part, the application of a mixture of POSS and copper phthalocyanine molecules as fillers for aminopropylsiloxane based composite material gives, in a properly chosen dopants concentration, homogenous films with an extremely low dielectric constant in the range of 1.8.

II. Experimental details

II.1. X-ray photoelectron spectroscopy

The photoelectric effect has been originally observed by Heinrich Hertz who produced the photoelectrons by illuminating an electrode by ultraviolet light. His work has been reported in a paper to the "Annalen der Physik" in 1887 [16]. This phenomenon was further outlined by Einstein in 1905 [17]. When the electromagnetic radiation is incident upon a surface, the interaction between the irradiating photons and the sample leads to the creation of photoelectrons (Fig. II.1.1a and II.1.1b). The kinetic energy, E_K, of the electrons emitted in this fashion equals:

$$E_K = hv - E_B - \Phi \quad \text{(II.1.1)}$$

where Φ is the work function of the sample, hv is the photon energy and E_B is the binding energy of the initial state of the electron [18].

Based on the photoelectric effect X-ray photoelectron spectroscopy (XPS) has been developed by K. Siegbahn and his research group. Around 1950's they decided to apply previously gained experience of the nuclear spectroscopy to the photoelectric effect. In 1954 the first photoelectron spectrum of the sodium chloride has been obtained. Afterwards a measurable binding energy shift of the photoemission peak depending of the chemical environment has been found [19]. XPS became an important method for establishing the chemical state of elements within the solid surfaces and K. Siegbahn was awarded the Nobel Prize for Physics in 1981 for its development.

During the XPS measurements a solid sample is mounted on a sample holder and is exposed to a source of the soft X-rays. For XPS, Al Kα (1486.6 eV) or Mg Kα (1253.6 eV) are often the photon energies of choice. The application of the monochromatized X-ray source gives an opportunity for the improving of the typical XPS resolution from 1 eV to 0.5 eV [18]. Ejected electrons are energy filtered via an analyzer and then the intensity for a defined energy is recorded by a detector. As a result photoelectron peaks reflect the different orbital energies of the atoms within the sample surface. The binding energies of the peaks are characteristic for each element. The elemental sensitivity and accuracy of this technique are about 1 % [18]. The peak areas can be used (while applying the appropriate sensitivity factors) to quantity the analyzed surface chemical composition.

a)

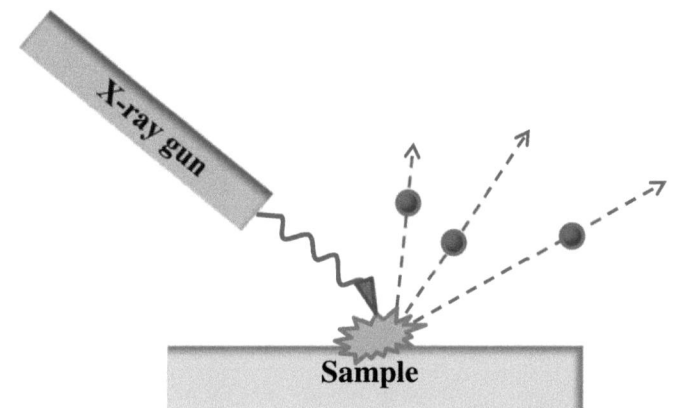

b)

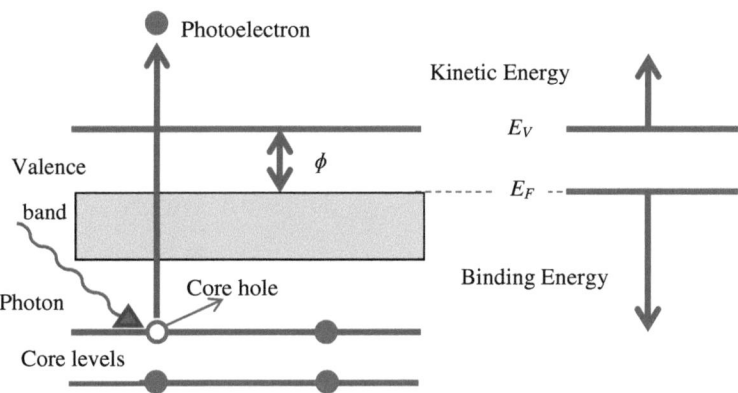

Fig. II.1.1. The scheme of the photoemission process.

Interestingly, not all peaks observed in the XPS spectra originate from the ejection of the electron due to the direct interaction with the photon. The occurrence of the Auger features is assigned to the relaxation of a higher energy electron to the state of the lower energy, possessing a vacant hole created by the X-ray photon. This process, illustrated in Fig. II.1.2, is accompanied by the emission of another electron of the energy equaling to the difference between the states involved in this phenomenon.

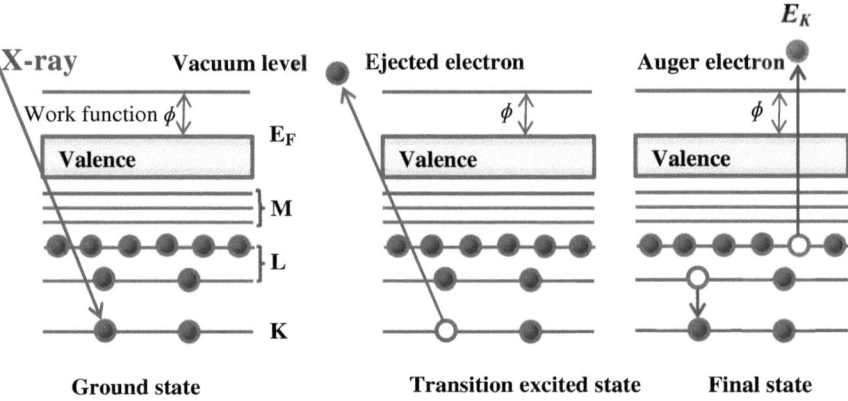

Fig. II.1.2. X-ray induced Auger electron.

The chemical environment of the surface atoms results in well-defined energy shifts of the peaks positions (ΔE). For two different compounds A and B, ΔE of a particular core-level has been described by the following equation [18]:

$$\Delta E(A, B) = K_c (q_A - q_B) + (V_A - V_B) \quad (II.1.2)$$

Where K_c is a coupling constant, q_A and q_B are the HOMO (highest occupied molecular orbital) energy level charges and the second term of the equation has a character of the Madelung potential arising from all the other ionic charges q_j centered at positions R_{ij} relative to the atom, i, in the material [18]:

$$V_i = \sum_{i \neq j} \frac{q_j}{R_{ij}} \quad (II.1.3)$$

In general, after the summation of the two terms of equation (II.1.3) for solids a value of a few eV or less is obtained for the chemical shifts in solids.

Since the escape depth of the ejected electron is limited, for energies around 1400 eV those located deeper than 10 nm below the surface undergo most probably energy loss event, contributing more to the background than to the primary photoelectric peak. This is the result of strong interactions of the photoexcited electron with the rest of the solid and may be expressed by the value of the mean free path λ, depending strongly on the photoelectron kinetic energy and related to the distance that the electron may overcome without an inelastic scattering [18]. The following equation describes the attenuation of the photoelectrons intensity after passing through the layer of the thickness d [18]:

$$I = I_0 exp\frac{-d}{\lambda} \quad (II.1.4)$$

where I and I_0 express the original and attenuated intensity.

The X-ray photoelectron spectroscopy (XPS) measurements described in this thesis were done using SPECS GmbH X-Ray source (Al Kα 1486.6 eV or Mg Kα 1253.6 eV) and an energy analyser made by Leybold-Heraeus. The structure due to the satellite radiation has been subtracted from the spectra before the data fitting [20,21]. The peak intensities were acquired after Shirley background removal [22]. The obtained spectra were deconvoluted with Gaussian lineshapes used to describe individual components of the peaks in order to obtain detailed information about the nature of the chemical interactions [23,24]. Each peak was fitted using Levenberg-Marquardt least-square algorithm. Data for percent atomic composition and atomic ratios were calculated using the atomic and instrument sensitivity factors [25].

II.2. NEXAFS spectroscopy

The development of the sources of the broadband synchrotron radiation made X-ray absorption spectroscopy a widely used technique of the materials structure investigation in many fields of the natural sciences. The near edge X-ray absorption fine structure spectroscopy (NEXAFS) can provide specific information concerning the unoccupied states and alows for instance to discriminate between sp^3 and sp^2 configurations. In order to record NEXAFS spectra, the sample is exposed to a tunable, monochromatized high flux of X-ray photons that excites an electron from an atomic core level to empty or partially occupied valence electronic states. Therefore NEXAFS requires the use of a synchrotron radiation source. The current induced by the photon beam changes as a function of excitation energy. The direct absorption technique consists of recording the total current of the emitted photoelectrons what gives the opportunity of the element-specific absorption coefficient determination. However, inner shell vacancy may relax by undergoing a transition from a higher energy, occupied shell, resulting in Auger electrons and fluorescence photons emission. Summing up, although the process accompanying the migration of the electron to the surface is still a matter of exploration and the general theory of the photoemission is complicated, one may simplify the considerations regarding the yield spectra and assume that the three following processes contribute to the photoelectric yield after the primary absorption of the photon:

- direct excitation to the conduction band
- emission of the Auger electron
- excitons decay

These three processes are illustrated in Fig.II.2.1. The determination of the electronic character of the unoccupied electronic states and the inter-atomic bindings is possible after monitoring the incoming I_0 and the transmitted flux when the photon energy is scanned through the absorption threshold [18,26–31].

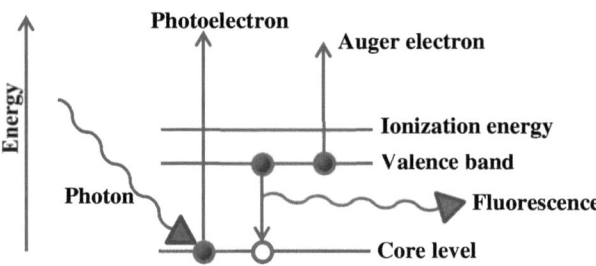

Fig. II.2.1. The processes contributing to the photoelectric yield after the primary absorption of a photon.

Summing up, the detection may be performed in TEY (total electron yield) mode where all photoelectrons leaving the sample, including direct and secondary, produced by Auger transitions, are monitored. However, in this situation the probing depth depends on the electrons attenuation length and for the C-K edge equals around 4 nm what makes TEY a surface sensitive mode. More bulk sensitive is the TFY (total fluorescence yield) mode where the spectra are recorded while monitoring the fluorescence radiation emitted when electrons of the lower binding energy fill the core holes generated by the absorption of the x-ray photons. The probing depth of this mode is given by the penetration depth of the incident and emitted photons and equals over 100 nm. TFY sums over all or a broad range of photon energies [30].

All NEXAFS spectra presented in this thesis have been performed using the BTU owned beam line U49/2 at Bessy II, Berlin [32]. The spectra were normalized to the X-ray intensity I_0, measured on a diode or Au mesh.

II.3. FTIR spectroscopy

Fourier transform infrared spectroscopy (FTIR) is an analysis technique providing information about the chemical bonding and molecular structure of materials. This technique is based on the fact, that the molecule exposed to the infrared radiation is excited to a higher energy state (mostly vibrational modes) while undergoing the quantized absorption process. The transition between quantized vibrational energy states results in infrared spectra. For molecule containing N atoms the degrees of freedom equals 3N, where six of them are related with the translational and rotational motion along the x, y and z axis and the remaining 3N-6 degrees of freedom correspond to the number of the vibrational modes in a nonlinear molecule. In many vibrational modes only a few atoms exhibit large displacements while the rest of the molecule is almost stationary. Different functional groups reveal a characteristic frequency of such modes where the motion is centered at a position being minimally affected by the nature of other atoms building the molecule. Therefore, the observation of spectral features allows their identification. So called fingerprint bands are very useful in order to distinguish one molecule from another containing similar functional groups. This bands involve the significant motion of only a few atoms and their frequency varies depending on the kind of molecule possessing a specific functional group [9,33,34].

For each mode, i, the atoms undergo an approximately harmonic displacement from their equilibrium positions and vibrate at a certain characteristic frequency, v_i. The vibrational energy states for the atoms vibrating with simple harmonic motions are given by the equation:

$$V_{iv} = hv_i \left(v_i + \frac{1}{2}\right) \quad \text{(II.3.1)}$$

where h is Planck's constant, v_i the fundamental frequency of the mode i, v_i the vibrational quantum number for the mode i (v_i =0, 1, 2, 3…). For the most vibrational modes the excitation from the ground state (v_i =0) to the first state of vibrational modes (v_i =1) requires an energy corresponding to a radiation in the mid-infrared spectrum (400 to 4000 cm^{-1}) [9,33,34].

The FTIR spectra presented in this thesis have been recorded using FTS 60B BioRad (with an exception of the chapter III.6, where the FTIR data were obtained and interpreted by the group from Wroclaw University of Technology).

II.4. AFM

Atomic force microscopy (AFM) is a mechanical profiling technique generating three-dimensional maps of the surface topography. After its invention in 1986 by Binning et al. AFM became a powerful and versatile technique for atomic and nanometer-scale characterization of a wide range of various surfaces [35]. The introduction of this investigation method opened a possibility of obtaining atomic resolution in the microscopy of conductors and insulators. Furthermore, this technique plays an important role in the fields of nanomanipulation, nanoassembly and molecular biology [35–37].

In AFM technique a sharp tip (force probe) attached to the end of the cantilever is affected by the attractive or repulsive forces generated by the interaction with the surface (Fig. II.4.1). The resulting deflection of the cantilever is detected mostly by the optical lever technique in which a laser beam is reflected from the cantilever and back onto a split photo-detector. The voltage signal produced by the photo-detector is used to control the tip-sample interaction force through the feedback loop mechanism [37].

AFM can be operated in several modes including [37]:

- contact mode (C-AFM) where the cantilever tip is located less than 1 nm above the surface,
- non-contact mode (NC-AFM) in which the cantilever tip vibrates several nanometers above the surface of a sample,
- tapping mode (TM-AFM) where the oscillating probe makes an intermittent contact with a sample (in this case the tip position above the surface is in between the contact and non-contact modes).

In general, the modes differ by the location of the AFM probe above the surface and thus by the pressure generated by the tip on the surface. The pressure in the contact mode is significant and lay in the GPa range and therefore cannot be used for soft samples as it may modify the surface during the measurement. As the investigated samples of this thesis were soft we used the non-contact technique in all the measurements which were conducted by Veeco CPII at room temperature. Phosphorus doped silicon cantilevers (model MPP-11123-10) with a resonant frequency of about 275 kHz and an approximate spring constant of 48 N/m were used. The back side of the cantilever was covered by 50 nm of aluminium. The nominal tip curvature of the used cantilever was less than 8 nm.

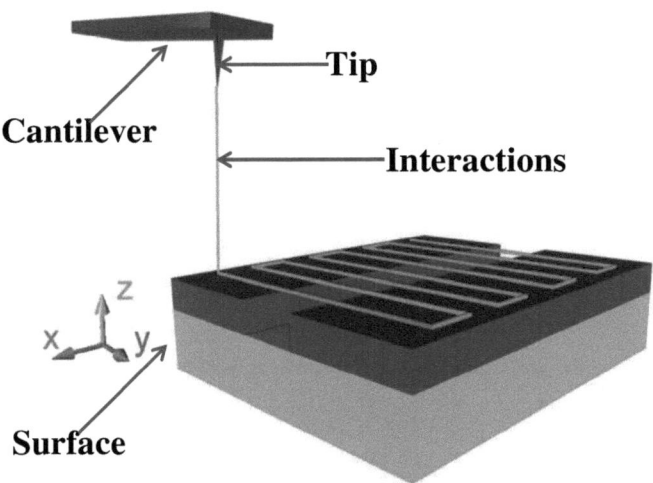

Fig. II.4.1. The idea of the AFM measurement.

II.5. Capacitance-voltage measurement

From several samples deposited on the silicon substrate described in this thesis metal-insulator-semiconductor (MIS) structures have been created by evaporation of silver contact on top of the spin-coated material. The dielectric constants of such prepared MIS structures have been estimated from the accumulation region of the capacitance-voltage characteristics.

In the CV measurements a differential capacitance is measured by a small (typically 20-35 mV) modulation of a DC bias voltage applied to a material stack. The DC voltage is biasing the semiconductor surface potential. In the equilibrium state of the MIS capacitor, the change of the gate charge is balanced by a change of the charge present in the semiconductor surface and interface. During CV measurements, the semiconductor is driven through accumulation, depletion and inversion. When a negative voltage is applied on the gate (in case of a p-type semiconductor), the holes accumulation at the semiconductor surface takes place. While driving the voltage towards positive direction, the Fermi level on the metal side is moving down. The hole charge density becomes lower and the total MIS capacitance is decreasing. The further increasing of the positive voltage leads to an additional increase of the space charge region (deep depletion). When the electron concentration at the surface equals the hole concentration in the bulk, a strong inversion is achieved. The large amount of the inversion carriers screens the further increase of the electric field. Thus the application of a higher positive voltage than that corresponding to the inversion conditions will not change the space charge region anymore. CV measurements may provide several information concerning the MIS stack, for instance: doping type of the semiconductor, the equivalent thickness of the insulator, the k-value of the insulator, the doping concentration [38].

The CV curves discussed in this thesis have been recorded on MIS structures by a CV set-up based on a LCR meter Agilent 4284A [39]. A slow ramp of 25 mV/s was used for the DC bias, which was superposed by a 1 MHz AC signal of 25 mV (RMS). The samples were first driven into the accumulation state, followed by biasing the semiconductor through depletion and inversion and then backwards to accumulation. The permittivity of the investigated layers was determined using the accumulation capacitance via the plate capacitor approximation as described below.

II.6. Deposition techniques

The samples described in this thesis consisted mostly of layers deposited on Mo or Si substrates. Mo has been used in the first stage of the investigation of the APTMS based samples in order to estimate the feasibility of the coating process. It has been assumed that an evidence of APTMS deposition onto Si(001) substrate may be confused with the presence of SiO_2 native oxide when XPS technique is utilized. However in most cases the silicon substrate has been used in order to enable the building of MIS structures for CV measurements on the produced samples. An important motivation favoring the Si substrate is also its wide applications in the field of microelectronics devices.

For the layers preparation molecules being a reach source of carbon species have been chosen. Besides commercially available substances (like C_{60}, PCBM, CuPc, POSS) for some films fabrication fullerenes encapsulated with silica shells through the micro emulsion synthesis prepared in our laboratory have been used. In order to perform this synthesis 10 ml of C_{60} in cyclohexane saturated solution, 1,3ml of Synperonic Np-5, 400 µl of C_{60} in toluene saturated solution and 80 µl of TEOS (tetraethyl orthosilicate) were added in a flask under vigorous stirring. Thirty minutes later 150 µl of ammonia aqueous solution was introduced and the mixture was stirred during the next 24

h. The substance obtained in such a way has been used for the preparation of samples described in the chapter III.1.

The simplest methods of the material deposition were dropping of the suspension of the investigated powders (mostly C_{60} or CuPc) or spraying of the unsaturated solution of the chosen molecules onto the substrate. These techniques have been applied in the preliminary stage of the experiments in order to estimate the influence of the particular factors (like solvents, exposition to the light) on the stability of the investigated substance.

Evaporation is a valuable *in situ* method which allows the obtaining of a high quality pure C_{60} layer. This technique has been used for the preparation of the reference fullerene samples and hybrid organic-inorganic composite materials as well (see chapter III.5). In order to choose the evaporation parameters that could be adapted in our experimental conditions, a series of experiments have been performed. As revealed by the investigation, the temperature of 800 °C was the most proper in order to perform this process effectively. The accompanying pressure was around 1×10^{-5} mbar. Fig. II.6.1 illustrates the thickness evolution of a C_{60} layer deposited onto a pure silicon substrate as a function of the evaporation time. These data have been obtained as a result of the calibration performed at the beginning of the experiments. The thicknesses of the fullerene layers have been calculated basing on the Eq. II.1.4 and taking into account the attenuation of the intensities of the XPS Si 2p core level signals.

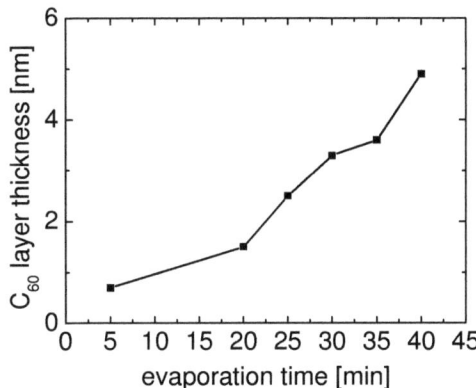

Fig. II.6.1. C_{60} layer thickness as a function of the evaporation time performed in our experimental conditions.

Here one should point out, that the coating of the substrate with a thin layer of APTMS significantly accelerated the growth of the fullerene layer produced by evaporation (see chapter III.5).

Fig. II.6.2. illustrates another exemplary experiment concerning the evolution of the C1s core level with time performed on the Mo surface coated with SiO_2 synthesized in our laboratory by means of the sol-gel method.

Some samples described in this thesis were deposited by means of the spin-coating. This process (illustrated in Fig. II.6.3.) involves a dropping of a small amount of a fluid onto the center of a substrate followed by the spinning of the substrate at a high speed. The spin coated fluid consisted in general of the molecules being the reach sources of the carbon species either dispersed or dissolved within the continuous medium. The reasons for the specified components application are given in the particular chapters of the thesis. In general, the priority in selecting a suitable continuous phase was to obtain a proper viscosity of the fluid to be used for the spin coating which at the same time ought realize the sufficient adhesion to the surface. For instance several films, mentioned briefly in the chapter III.1, were produced by the addition of the polyethylene glycol as a thickener to the fullerene toluene solutions, in other cases the reaction mixture functioned as a continuous phase containing C_{60} encapsulated with silica shells.

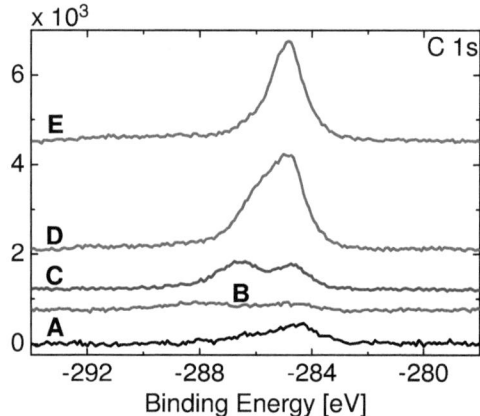

Fig. II.6.2. C 1s core level XPS spectra obtained for: **A)** pure Mo substrate, **B)** Mo substrate coated with the SiO_2 by means of the sol-gel method, **C)** surface B after 10 minutes of the C_{60} evaporation in 800°C, **D)** surface B after 35 minutes of the C_{60} evaporation in 800°C, **E)** surface B after 60 minutes of the C_{60} evaporation in 800°C.

As a continuous medium for the films investigated by CV technique the APTMS monomer has been used. In order to shorten the time of the experiments and to obtain reliable information concerning the possible interactions between the components itself (mainly by XPS investigation), neither external catalyst nor solvent has been added during the preparation routine of the composite APTMS based films.

The silicon substrates used in the spin-coating procedure were mostly treated with the piranha solution (a mixture of 7 : 3 (v/v) 98 % H_2SO_4 and 30 % H_2O_2), since the hydroxyl groups formed in this manner improve the homogenous deposition of the APTMS solution through the reactions with the methoxy groups.

A series of the spin-coated films have been produced while using a wide range of the rotational speed (500-6000 rpm). Regarding the APTMS based fluids the films formed at 6000 rpm were the most proper to perform AFM and CV measurements, thus this speed was mostly applied in the experiments described in this thesis.

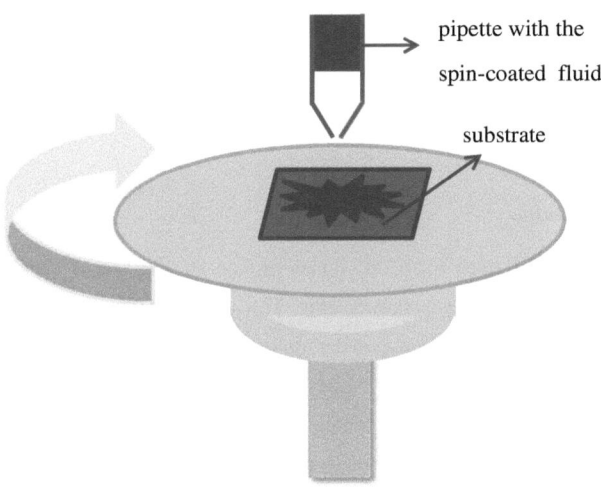

Fig. II.6.3. The scheme of the spin-coating technique.

III. Results and discussion

III.1. The stability of C_{60} and its derivatives upon handling in microsystems technologies[1]

III.1.1. Introduction

Fullerene science is one of the fastest growing areas of research in chemistry, physics and material science [40]. As well-known and basic nanomaterials, fullerene and their derivatives with tunable shape and size have attracted great attention for their potential applications in, for example, optoelectronic devices [41–43]. Much attention has been paid to C_{60}, one of the fullerene family, because of its potential applicability to advanced materials [44–46]. This chapter is an overview of the first attempts leading to the preparation of the fullerene containing layers and contains the general results of the spectroscopic investigations. Selected sorts of the samples are described in more details in further chapters of this thesis.

III.1.2. Experimental

III.1.2.1. Materials

Fullerene C_{60} (99,5 %), toluene (99,8 %), cyclohexane (99,5 %), chlorobenzene (99,5 %), ethanol (99,5 %), 3-aminopropyltrimethoxysilane (97 %), ethylene glycol were obtained from Sigma Aldrich. Fullerol has been received from AMD (Dresden). Fluorinated fullerenes were delivered by Moscow State University.

III.1.2.2. Samples preparation

Fullerene and it's derivatives films on silicon surface were prepared by using following methods of covering: dropping, spraying, spin-coating. The sample of pure C_{60} used as a reference was obtained by evaporation of fullerene powder onto a silicon surface at 800°C in vacuum conditions. Spraying and dropping techniques were adopted in the case of fullerene dissolved in organic solvents with different polarity: hexane, toluene, cyclohexane. Solutions of fullerene derivatives were also sprayed and dropped: water solution of $C_{60}(OH)_{24}$ and ethanol solution of $C_{60}(F_3)_{12}$. In order to prepare film by spin-coating additional substances have been used: APTMS and polyethylene glycol. These substances have decreased viscosity of the fullerene solutions what allowed to obtain fullerene containing films. As a result several kinds of films have been prepared,

---

[1] J. Klocek, D. Friedrich, D. Schmeisser, M. Hecker, E. Zschech, The stability of C_{60} and its derivatives upon handling in microsystems technologies, in: Students and Young Scientists Workshop" Photonics and Microsystems", 2009 International, 2010: pp. 43–46.

including the fullerene encapsulated with silica shells. Films containing this substance were prepared by spin-coating. Table III.1.1 overviews the investigated substances and the spectroscopic measurements done of them.

III.1.3. Results and discussion

III.1.3.1. NEXAFS measurements

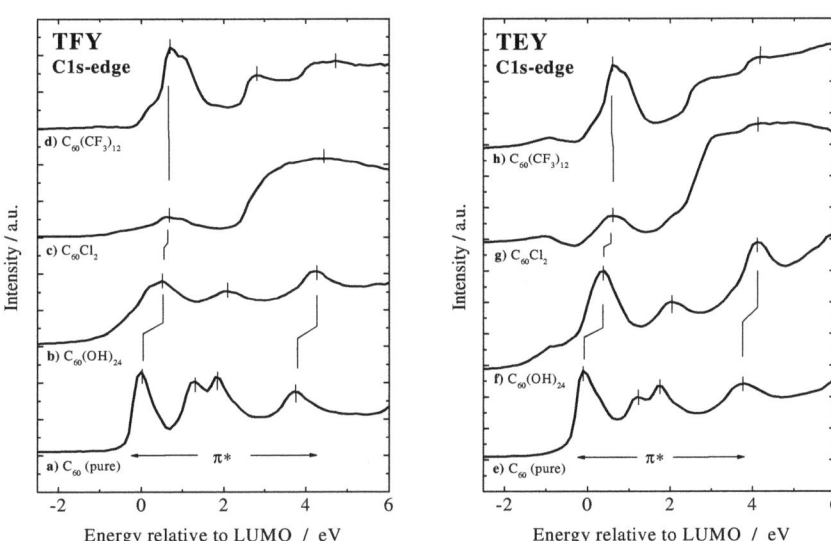

Fig. III.1.1. NEXAFS (TFY and TEY) spectra of C_{60} and some derivates. a, e) Pure C_{60} as reference, b, f) Fullerol, c, g) C_{60} with chlorine and d, h) C_{60} with fluorine.

NEXAFS spectra were recorded around the C 1s, O 1s, F 1s and Cl 2p absorption edge. The spectra at the C 1s edge are given in Fig.III.1.1. The C_{60} molecule with and without substitutional groups show clear differences in the XAS spectra: the K-shell emission around 284eV-290eV is attributed to the excitations from C 1s core level into empty states of the π^*-electron-system. The lowest unoccupied molecular orbital (LUMO) appears at around 284.3 eV. Further results are referred in relation to this energy level. The LUMO level corresponds to the t_{1u}- and t_{1g}-state of the C_{60} π^*-system. The peak located at 1.2 eV is related to the h_g-, h_u- and t_{2u}-states, while the peaks of the gg-state appears at 1.7 eV. The peak at 3.8 eV is correlated to the g_u- and t_g-state [47,48].

Based on the fact that C_{60} is a complex π*-electron-system, it is possible to characterize the stability of the "soccer ball" structure of C_{60}. Therefore one may draw some conclusions concerning the influence of the substitutional groups of the π*-electron-system and thus of the C_{60} molecule's structure stability. In the NEXAFS spectra of fullerol ($C_{60}(OH)_{24}$) a clear shift of the peaks as a consequence of the electro negativity of the bonded OH-groups is observed. This chemical shift equals about 0.5 eV. The spectra are very similar to the pure C_{60} profile. Basing on this fact one may conclude that the molecule is stable. There is also observed a chemical shift in the spectra of C_{60} with fluorine and chlorine. It is about 0.6 eV (-CF_3) and 0.7 eV (-Cl_2), but the spectra of these two materials show strong changes. Therefore, the molecules shape is probably changed.

III.1.3.2. FTIR measurements

In every FTIR spectra of the films obtained by spraying and dropping of a fullerene solutions, as well as of samples that contain additionally polyethylene glycol, there are observed four peaks, approximately at: 525, 575, 1182 and 1429 cm^{-1}. These values are corresponding to T_{1u} vibrations of the C_{60} molecule [49,50]. These four peaks do not occur in the case of FTIR spectra of the samples prepared by mixing of the fullerene with APTMS. While analyzing spectra of the samples containing fullerene and APTMS one observe a probable chemical reaction between the nucleophilic nitrogen atom from APTMS and electrophile fullerene molecules. In contradiction clear peaks around 525, 575, 1182 and 1429 cm^{-1} are noticed when APTMS is only used to modify the surface before the spin-coating of the fullerene solution.

III.1.3.3. Raman and XPS measurements

Raman measurements were performed in a backscattering setup on a Horiba Yobin Ivon LabRAM HR spectrometer using a 488 nm Ar-ion laser excitation. The Raman spectra of films obtained by spraying of a fullerene toluene solution contain three peaks around: 1425, 1470 and 1575 cm^{-1}. These values are corresponding to the frequencies of Hg vibrational modes of a C_{60} molecule [51]. The XPS data of fullerene layers prepared by spraying and dropping contain a strong signal around -285 eV, including sp^2 and sp^3 carbon atoms contribution. More detailed interpretation of the XPS spectra of the spray-coated layers has been performed in chapter III.2.

Table III.1.1. Spectroscopic investigation of fullerene and its derivatives stability within films obtained upon processing: spraying, dropping, spin-coating, evaporation. n.i means: not investigated.

film composition	coating method	presence of fullerene characteristic features			
		FTIR	NEXAFS	XPS	RAMAN
C_{60}	evaporating	n.i.	yes	n.i.	yes
$C_{60}(F_3)_{12}$, C_2H_5OH	spraying	n.i.	yes	n.i.	yes
$C_{60}(OH)_{24}$, H_2O	dropping	n.i.	yes	n.i.	yes
$C_{60}Cl_2$	dropping	n.i.	no	n.i.	no
C_{60}, $C_6H_5CH_3$	spraying	yes	no	yes	yes
C_{60}, $C_6H_5CH_3$ $HN_2(CH_2)_3Si(OCH_3)_3$	spin-coating	yes	no	n.i.	n.i.
C_{60} with SiO_2 shells, $C_6H_5CH_3$, $C_{2n+2}H_{4n+6}O_{n+2}$	spin-coating	yes	no	n.i.	n.i.
C_{60}, C_6H_5Cl $HN_2(CH_2)_3Si(OCH_3)_3$	spin-coating	no	n.i.	n.i.	n.i.

III.1.4. Conclusions

The results of spectroscopic investigations confirm the stability of fullerene molecules within the films obtained by fullerene powder evaporation in vacuum conditions. After the detailed analysis of the results of obtained films one conclude, that the technique of the C_{60}-containing films preparation including evaporation of fullerene molecules should significantly improve the stability of the new material. However, this method can be used only in order to obtain pure C_{60} films. In the case of fullerene derivatives spray-coating is an effective, non-destructive method of thin film preparation. Here, the important requirement is a proper choice of the solvent. Nonetheless, Raman spectroscopy suggests that the homogeneity of the films obtained by dropping and spin-coating is low. Based on spectroscopic data we observe that $C_{60}(F_3)_{12}$ is stable in ethanol. The stability of $C_{60}(OH)_{24}$ in water is also high. FTIR, Raman and XPS measurements suggest the presence of fullerene molecules in films obtained by spray-coating. However, NEXAFS excludes the stability of C_{60} molecules in these layers. These results suggest, that basic spectroscopy methods can be used only in order to estimate the stability of fullerene molecules, the obtained data should be however confirmed by more sensitive spectroscopic methods, for instance NEXAFS. FTIR investigation suggests that there occurs a chemical reaction between APTMS and fullerene molecules. In opposite, the FTIR data contain four peaks that are characteristic for fullerene molecules, when APTMS is used only for the pretreatment and modification of the surface in order to increase its adhesion, and the C_{60} coating is carried out afterwards onto this modified substrate.

III.2. Toluene influence on the thin fullerene and copper phthalocyanine film properties

III.2.1. Introduction

Thin films composed of copper phthalocyanine (CuPc) or fullerene molecules possess besides other interesting properties also unique chemical and thermal stability. Hence, these films are an object of considerable interest because of their potential applications in many fundamental technologies related to coating and surface modification ranging from devices for solar energy conversion, biosensing to advanced nanostructured devices for optics and microelectronics [52–56]. Due to its special properties these compounds may be also applied as guest molecules within hybrid nanocomposites that are presently of great interest in the nanotechnology [57], giving the opportunity to transfer its outstanding physical and chemical properties into the hybrid material. CuPc oligomers found among many others an application as a high dielectric constant filler in a polymer matrix [58,59]. On the other hand, great effort has been made also in order to obtain solid-state fullerene containing nanocomposite materials [54,60]. According to theoretical calculations the proper arragment of the fullerene molecules may significantly decrease the value of the material dielectric constant [61]. In addidion, the presence of fullerene cages within the hybrid composite can considerably improve its mechanical properties [54,60].

Although CuPc and fullerene containing films open a promising horizon in many fields of science, the preparation techniques may affect the desired properties of the final films due to such factors as the solvent application (required for instance in Sol-Gel method). The present chapter is composed of two series of investigations. In the first part the influence of the toluene on the CuPc and C_{60} chemical properties has been measured by means of X-ray photoelectron spectroscopy (XPS). I also try to estimate whether the detectable interaction between these two compounds takes place in toluene environment. Second part of the chapter is related to the spectroscopic investigation of the C_{60} layer obtained by spraying its toluene solution onto silicon substrates.

III.2.2. Experimental

III.2.2.1. Materials and substrates

C_{60} (99,5 %), PCBM (99,5 %), APTMS (97 %) and POSS, obtained from Sigma-Aldrich has been used. CuPc (95 %) has been delivered from Alfa Aesar. N-type silicon (100) wafers (Phosphorous doped, Crystec, Berlin, Germany) were chosen as substrates.

The silicon substrates were cleaned by ultrasonification in acetone, isopropanol and then in distilled water for 10 minutes and were dried afterwards in nitrogen stream. The thin layers of the samples (*A*, *B*, *C* and *D* in Fig. III.2.1) were obtained from dense suspension of CuPc and C_{60} powders or its mixture in toluene. The films were deposited on the silicon surface by dropping followed by solvent evaporation in room conditions.

For the film E preparation a 0.5 % C_{60} toluene solution has been sprayed onto the substrate. The reference sample F was produced by fullerene powder evaporation onto the Si surface at a temperature of 800°C and a pressure of 1×10^{-5} mbar for 120 minutes in order to obtain a thick film. The XPS spectrum of the evaporated C_{60} layer has been recorded in situ, directly after the deposition. The schemes of the samples A, B, C, D, E and F are presented in Fig. III.2.1.

III.2.3. Results and discussion

III.2.3.1. XPS investigation of the samples A, B, C and D

Here the XPS C 1s core level spectra obtained for the samples A, B, C and D will be discussed. The main purpose of this experiment was to estimate the influence of the toluene solvent on the chemical properties of the investigated substances and to inspect presumptive interactions between CuPc and C_{60} in the toluene environment. The C 1s peak, centred at 284.6 eV, was used to calibrate the spectra of the samples since this value of the binding energy is being used as a standard for both CuPc and C_{60} XPS spectra [62–64]. The C1s core level XPS spectra of the samples A, B, C, and D are presented in Fig. III.2.2, whereas the Table III.2.1 lists the binding energies extracted by the peak fitting procedure.

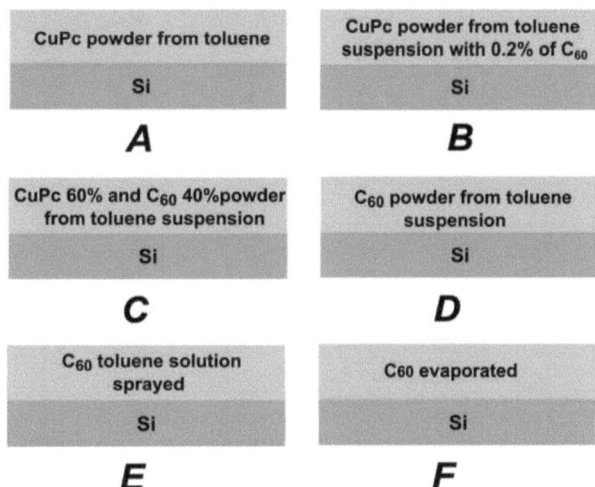

Fig. III.2.1. The schemes of samples with drop casted layers produced from toluene suspensions on Si substrates: *A)* CuPc powder (referred in the text as sample *A*), *B)* CuPc powder containing 0.2 % C_{60} (sample *B*), *C)* mixture of powders containing 60 % CuPc and 40 % C_{60} (sample *C*), *D)* C_{60}

powder (sample *D*), *E*) sample produced by spraying of the C_{60} toluene solution on the silicon substrate (sample *E*), *F*) reference sample produced by C_{60} evaporation on the silicon substrate (sample *F*).

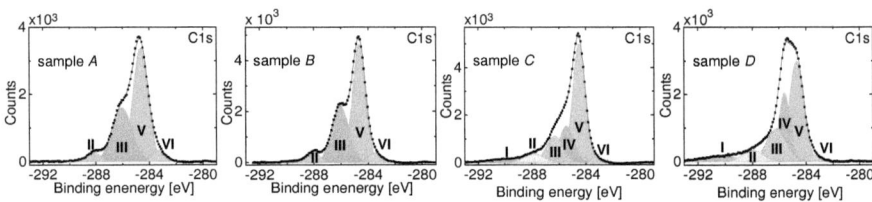

Fig. III.2.2. C 1s core level XPS spectra recorded for samples *A, B, C* and *D*.

Table III.2.1. C 1s XPS spectra decomposition of the samples *A, B, C* and *D* and its comparison to the literature data.

peak	Pure CuPc (literature [65])	Sample A	Sample B	Sample C	Sample D	Pure C_{60} (literature [64])	assignment
I				-290.2	-290.2	-289.8	C_{60} shake-up[63]
II	~ -288	-288.2	-287.9	-288	-287.9	-287.8	shake-up satellites of the aromatic carbons [66]/ oxygen-containing moieties[67] / C=O species [68]
III	-285.8	-286.0	-285.8	-286.4	-286.1	-286.1	C-O[68] /carbon C3 of CuPc molecule [69]
IV				-285.6	-285.6		sp3 carbon atoms[70,71]
V	-284.7	-284.6	-284.6	-284.6	-284.6	-284.6	C-C of fullerene cages[68,72]/ C of the CuPc benzene rings[69]
VI			-283.3	-283.3	-283.4	-283.3		Si-C[73]

24

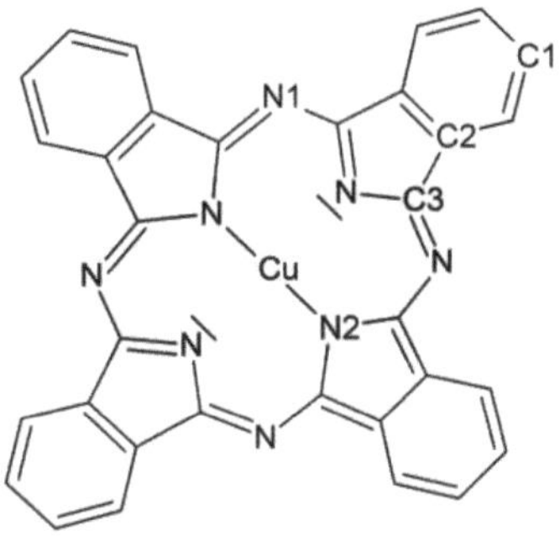

Fig. III.2.3. Scheme of the CuPc molecule.

While analyzing the decomposition of the C1s core level spectrum of the sample *A* one notice a good agreement with the chemical bonding states found in the literature [65,66,69]. Therefore, one may suppose that the solvent did not influence significantly the chemical properties of the CuPc.

The most pronounced feature in the XPS spectrum of the sample *A* (signal V) originates from the 24 atoms of the 4 benzene rings of CuPc [69] (including carbon C1 and C2 in Fig. III.2.3). The signal III is attributed to the 8 carbon atoms linked to nitrogen atoms [69] (marked as C3 in Fig. III.2.3). The difference between the center positions of the signals V and III equals 1.4 eV showing an excellent agreement with the literature data [69]. The nature of the feature II is somewhat controversial. Its central position at around 288 eV corresponds to the C=O group. However Niwa et al. [66] noticed the presence of this feature in the XPS spectra of aromatic compounds which are stable against oxidation. Thus they assigned it to shake-up satellites of the aromatic carbons. Nevertheless, taking into account that the XPS spectrum of the sample *D*, which is not consisting of any CuPc, also contains this feature, one cannot exclude the additional contribution of the C=O species to the C1s spectra of the samples *A*, *B* and *C*. The source of the weak signals VI present in all spectra described in this part and located in the range of the binding energy that corresponds to the Si–C species is also not quite clear. A similar feature is also reported in the literature concerning the XPS analysis of CuPc films [74] and has been in general classified as not belonging to the CuPc carbon component. Nevertheless, for the samples *A*, *B*, *C* and *D* one can attribute these features to the Si–C formation due to interactions between the deposited materials and the silicon substrates. Since the films were composed of the powders they could be discontinuous. Hence, the chemical composition of the substrate-film interfaces can also be detected.

While the toluene did not affect significantly the chemical composition of the CuPc powder, its influence on C_{60} can be clearly seen in the C 1s spectrum of the film D. As a result of the main feature decomposition one obtains not only the typical C_{60} cage signal at around 284.6 eV (peak V) but also a second component at 285.6 eV (peak IV). According to the literature, these binding energy positions are typical for sp^2 and sp^3 carbon contributions, respectively [71,75,76]. Interestingly, the C 1s core-level shift between the sp^2 and sp^3 contributions realized in the fitting process equals 1 eV and is in excellent agreement with the theoretical value obtained by means of molecular dynamics simulations [77]. The quantitative analysis reveals an atomic C:O ratio within the sample D of 5.7:1 (the O1s core levels are not shown). Taking into account, that the pure C_{60} molecule does not contain oxygen, one can classify the detected O species as a contamination formed as a by-product of the fullerene dispersing in the organic solvent. The fullerene oxidation explains at the same time the sp^3-hybridized carbon atoms formation. Thus one may presume that toluene accelerates the C_{60} oxidation in ambient conditions. Such an evident influence of the toluene on the chemical structure of C_{60} is a bit surprising taking into account that exactly this solvent is commonly used for the fullerene separation process [78]. However, changes in the C_{60} electronic structure after its treatment with the organic solvent have been already mentioned in the chapter III.2. Nevertheless the presence of the signal I, which is attributed to the characteristic shake-up feature of fullerene corresponding to π-π^* transitions, is the proof (besides the signal V) that only a part of the fullerene molecules present within the film D underwent the by-reactions resulting in sp^3 carbon formation. The source of the signal III in the sample D C1s spectrum are oxidized carbon species [67,68].

While analyzing the C 1s core level spectra of the samples B and C one cannot notice the evidence of the chemical interactions between the CuPc and the C_{60} molecules within the powder. The spectrum of the sample C confirms that mostly fullerenes are affected by the toluene resulting in the sp^3 carbon formation, since the feature IV is only present at a fairly high C_{60} concentration (40 % in sample C). The presence of the signal I in the C 1s spectrum of the sample C suggests further, that a large amount of the –C=C– fullerene species has not been affected by the by-reactions during the preparation process, similarly to pure C_{60} (sample D). Based on the comparison between the spectra of the samples A, B, C and D one may exclude a partial contamination of toluene as a source of the increased sp^3 contributions in the sample C and D since for the pure CuPc (A) and for the film with a very low C_{60} concentration (B) the feature IV does not exist. The fullerene concentration within the film B was too low to detect clearly the C_{60}-cage shake-up features in its C 1s core-level spectrum.

III.2.3.2. XPS investigation of the samples *E* and *F*

Since the results described above proved that the toluene affected especially the chemical properties of the fullerenes, in the next step of the experiment I produced a fullerene layer deposited by spraying of 0.5 % C_{60} toluene solution on the Si substrate (sample *E*). The C1s core level XPS spectra of the sample *E* and the reference sample *F* are presented in Fig. III.2.4.

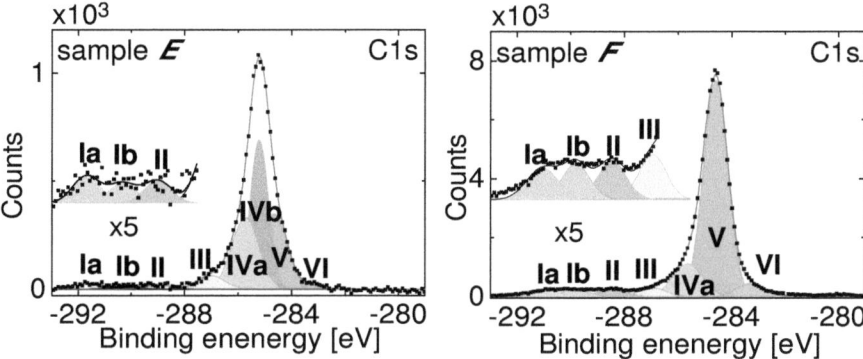

Fig. III.2.4. C 1s core level XPS spectra recorded for samples *E* and *F*.

The decomposition of the C 1s core level spectra revealed a significant decrease of the sp^2 hybridized carbon species within the film *E* in comparison to the film *F* (signals V). At the same time the sp^3 (signals IVa) related signals are behaving reversed. Simultaneously, the C 1s core-level shift between the sp^2 and sp^3 (signals IVa) contributions remains 1 eV for both samples, what is in agreement with the results obtained for the samples *C*, *D* and with the literature data [77]. Quite interesting is the appearance of the pronounced additional feature IVb in the sample *E* spectrum. According to the literature data, the binding energy of 285.1 eV, corresponding to the center position of the signal IVb, has been observed in C 1s XPS spectra of aromatic hydrocarbons like benzene [79]. The presence of this feature was also revealed by NEXAFS investigations (described below). Since the amount of the solvent molecules potentially adsorbed on the fullerenes seems to be too low to result in such a pronounced feature one may presume, that the dissolving of C_{60} in toluene under ambient conditions partially damaged the structure of the fullerene cages, particularly the exposition of the solution to the light might be important. The formation of aromatic groups as one, among others, product of the fullerene decomposition has already been reported in the literature [80]. An additional presence of atmospheric impurities cannot be excluded too, since one should take into account that the sample *E* has been deposited in ambient conditions and exposed several days to the atmosphere before the analysis. In comparison, the peak IVb doesn't appear in the sample *F* spectrum which was recorded *in-situ* just after the evaporation in vacuum conditions. The percentage of the feature III peak area corresponding to the C–O species [71] within the overall C 1s signal is significant higher for the sample *E* than for the sample *F* what is the next consequence of the ambient influence. However, the presence of weak characteristic fullerene features attributed to the π-π* electron transition around 290 eV suggests, that not every fullerene molecule within film *E* has been affected by the interactions with the environment. However, the low intensities of the features Ia, Ib and II in the C 1s core level spectrum of the sample *E* hindered a precise decomposition process that for this part of spectrum *E* has been performed only indicative.

While analyzing the C 1s core level spectra of the samples *E* and *F* one may notice, that the source of the feature VI which center position corresponds to the binding energy of the Si–C binding energy is somewhat controversial. As for the samples *A*, *B*, *C*, and *D* the presence of this weak feature may be explained by the material discontinuity on the silicon substrate, for the thick evaporated C_{60} layer (sample *F*) it is a bit surprising. Therefore a marginal attention is paid to this feature. Table III.2.2 lists the peak decomposition results of the C 1s core level spectra of the samples *E* and *F*.

Table III.2.2. C 1s XPS spectra decomposition of the samples *E* and *F*.

peak	sample *E*	sample *F*	Assignments
Ia	291.6	291.1	C1s shake up satellite[63,81]
Ib	290.3	289.8	C_{60} shake-up [63,81]
II	288.9	288.4	C=O [82]
III	286.9	286.9	C–O [71]
IVa	285.6	285.6	sp^3 carbon atoms[70,71]
IVb	285.1	----	C–C, C–H [79]
V	284.6	284.6	C–C of fullerene cages[68,72]
VI	283.1	283.3	Si–C [73]

III.2.3.3. NEXAFS investigations of the sample *E*

The sample *E* has been additionally characterized by NEXAFS spectroscopy. In order to draw a conclusion concerning the influence of the toluene and ambient environment on the fullerenes present within the film E, the spectra have been decomposed and additionally compared with those obtained for the reference film F. Absorption spectra were recorded using the TEY and TFY methods. In TEY mode the total current leaving the sample is proportional to the absorption spectra. The TFY was measured with a fluorescence detector. The deconvolution of the NEXAFS spectra was performed in WinXAFS software using Gauss and asymmetric P-Voigt functions for the resonance peaks and an arctangent for a step function [83–85].

Fig. III.2.5 shows the C-K edge NEXAFS spectra of this sample recorded in the surface sensitive TEY and bulk sensitive TFY modes in comparison to the reference film F.

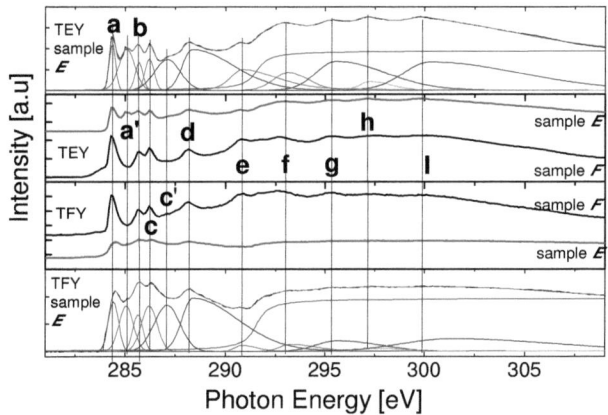

Fig. III.2.5. An exemplary decomposition way of the C-K edge NEXAFS spectra of sample E recorded in the TEY mode (upper part) and TFY (bottom part). The middle part represents the comparison of the spectra obtained for the sample E and for the reference sample F.

The fitting procedure of the particular resonances and steps in molecular K-shell spectra is an object of ongoing researches and explorations, both in theoretical and experimental fields [28]. The experimental examination of step-like features included in the near-edge spectra, that are a result of core electron excitation to the continuum of the final states, is also difficult since they are influenced by other signals [84]. Thus the main attempt of the deconvolution procedure performed in this chapter was to estimate the central position of the most pronounced and prominent resonances using preferably a low number of the functions.

The C-K edge NEXAFS spectrum of the sample E recorded in TFY mode exhibits less prominent features then that obtained in the TEY mode. This is justified since the X-ray fluorescence process is weaker than the non-radiative channels [86]. Nevertheless this fact seriously hindered the identification of the C 1s-σ* resonances positions for the TFY spectra. Therefore, in the following discussion a marginal attention is paid to these features.

The localized C 1s-π* transitions appear in a form of sharp features below the C1s ionization potential. In Fig. III.2.5 one may also observe some broader features corresponding to the C 1s-σ* resonances and finally a continuum part at the higher photon energy. The position of the carbon absorption peaks depend on the chemical environment of the carbon atoms [87], thus more detailed local bonding information may be obtained while comparing the center positions of these resonances with existing literature data.

The discrete part of the spectra recorded for the reference C_{60} sample *F* exhibits 3 intensive resonances localised at around 284.4 eV (signals a), 285.7 eV (signals b), and 286.2 eV (signals c), corresponding to the C 1s-π*$_{C=C}$ transitions [87–89]. These features are also found for the sample *E*. However in Fig. III.2.5 one may clearly observe that the spectrum of the sample *E* contains an additional feature (a') that is very pronounced in particular in the spectra of the surface sensitive TEY mode. The feature a' occurs at around 285.1 eV and has been assigned to the C1s(C–H)-π*$_{C=C}$ transitions [85,90]. This is in agreement with the XPS results described above and allows the presumption, that the fullerene film deposited by spraying contains aromatic hydrocarbons. The second prominent feature present in the TEY spectrum of the sample *E* (signed as c') appears at around 287 eV. This broad feature contains probably the contribution from both C 1s-π*$_{C=O}$ [91] and C 1s-σ*$_{C-H}$ [92] resonances. The feature d at around 288.3 eV existing in all spectra illustrated in Fig. III.2.5 is characteristic for π* resonances [87] and might be attributed to the C 1s-π*$_{R-(C=O)-R}$ transition [93] of the partially oxygenated fullerene molecule. At this point one should underline that the NEXAFS investigation has not been performed in situ and also the reference sample *F* has been stored in ambient conditions before this analysis, thus its oxidation was possible. The C 1s-π*$_{C=O}$ transitions probably contribute to the signals e located around 290.6 eV [88]. The peaks situated above the ionization edge around 293 (f), 295.5 (g), 297 (h) and 299 (l) eV originate from the C 1s-σ* transitions [88].

First of all, while comparing the C-K edge NEXAFS spectra of the sample *E* one observe a damping of the two first π* states indicating a partial re-hybridisation of the –C=C– species into C–C single bonds [94]. One of the possible explanations is the present of a larger number of C–O bonds within the sample *E* what revealed the corresponding XPS spectrum (Fig. III.2.4). According to the literature, these kind of oxygen functionalities emerging at the empty coordination sites are typical for the defective, holey molecules [94]. This damage of the fullerene cages can be however eliminated by thermal treatment which leads to the removal of the C–O functionalities, the creation of empty carbon coordination sites and its further saturation by the homo-polymerisation [94]. The appearance of the new localised double-bond feature assigned, similarly as it has been done for the corresponding XPS results, to the aromatic C–H species, may be a result of a specific photochemical reaction. The occurrence of the fullerene photochemical degradation in the organic solvent environment after UV irradiation, leading to the formation of the whole range of the C_{60} derivatives, has been already reported in the literature [80]. Nevertheless, the slow progress of this kind of reaction has also been observed to some extend when C_{60} benzene solution was exposed to the light without irradiation [80].

Summing up, NEXAFS investigations revealed that fullerene exposure to the air leads to the partial breakdown of some fullerene molecules and the formation of additional functionalities like carbonyl or carboxyl/ester groups. The decomposition of the C-K edge NEXAFS spectra revealed the existence of resonances originating from oxygenated and partially unsaturated system of the carbon-carbon bonds within the film E deposited by spraying.

III.2.3.4. FTIR investigation of the samples A and E

In order to confirm the presence of the not damaged C_{60} cages within the solvent and light treated fullerene layers, the samples A and E have been investigated by FTIR spectroscopy. Sample A has been produced in two series: from toluene suspension, referred further in text as sample $A_{toluene}$ and from hexane suspension, referred in text as sample A_{hexane}.

Although XPS and NEXAFS revealed the presence of a variety of carbon functionalities within the solvent-treated fullerene layers, FTIR exhibits that a large number of fullerene molecules remained intact. This is proved by the presence of the four allowed vibration bands of C_{60} fullerene [95,96] in the FTIR spectra observed for all films described in this part (features A around 525 cm^{-1}, B around 576 cm^{-1}, C around 1182 cm^{-1} and D around 1429 cm^{-1}).

The spray deposited film E was significantly thinner than the layers A thus the observed FTIR spectra vary in intensity. There is no noticeable change in the spectra obtained for the samples $A_{toluene}$ and A_{hexane}. The large number of weaker features in the FTIR spectra of these layers may arise from impurities previously present within the fullerene powder and further adsorbed from the air and some by-products of the interactions with the environment. Some of them may be classified based on the literature data. For instance, some contribution to the signals appearing at around 669-671 cm^{-1} may originate from the Si–O–Si bands [97]. The presence of the FTIR resonances in the range of 700-800 cm^{-1} has been observed in the spectra of, for instance, phthalocyanines and has been assigned to the out-of-plane C–H wagging [98]. The quite pronounced features observable at around 960 cm^{-1} may be attributed to the stretching of non-bridging oxygen atoms [99]. The signals around 1110 cm^{-1} may arise from the anti-symmetrical ring stretching [100]. The feature around

1164 cm^{-1} causing the broadening of the peak C may originate from the C–H in-plane bending mode observed also for the di-substituted benzenes [101]. Since the same feature is observed for layers obtained from both toluene and hexane suspension, the solvent molecules can be excluded as a potential source of these FTIR bands.

Although the proper assignment of all signals existing in Fig. III.2.6 is a quite challenging task, in general one may conclude that the features observed in the FTIR spectra of the samples A are, apart from the C_{60} characteristic resonances, generated by a considerable extent by the CH_2 deformations.

The C_{60} toluene solution used for the spraying was sediment-free thus the fullerenes deposited onto silicon by means of this technique were significantly purified in comparison to those deposited from the toluene suspensions. This could be another reason, apart from the lower intensity of the sample E overall FTIR spectrum, for the less number of observable additional features.

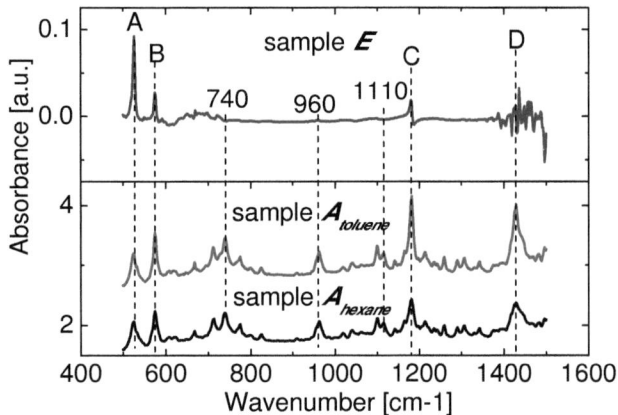

Fig. III.2.6. FTIR spectra obtained for the samples E, $A_{toluene}$ and A_{hexane}

III.2.4. Conclusions

The XPS results confirmed that the toluene does not affect the CuPc structure to a considerable extend. The spectra of the CuPc powder obtained from the toluene suspension and exposed to the air are very similar to those already published in the existing literature, produced by the evaporation in vacuum conditions [65]. On the other hand, the spectroscopic data of the toluene treated fullerenes revealed the significant increase of the sp^3 hybridized carbon species in comparison to C_{60} films deposited by sublimation. The decomposition of the C 1s XPS spectra of the toluene treated fullerenes confirmed the presence of the carbon atoms attached to different oxygen-containing moieties, with the superiority of the C–O species. Nevertheless the collected data confirmed that a relatively large fraction of C_{60} did not participate in the reaction with the environment. This could be explained by the kinetic limitation leading to the saturation of the interaction processes due to the formation of a passivation layer between the unreacted C_{60} and the environment [102]. The NEXAFS data of the spray deposited film E revealed a damping of the two leading π* resonances and the existence of additional prominent features corresponding to the C1s-π*$_{C=O}$ and C 1s-σ*$_{C-H}$ transitions that are not observable for the reference fullerene sample produced by fullerene evaporation and sublimation. This observation suggests that toluene increases the fullerene reactivity accelerating the formation of carbonyl functionalities and aromatic C–H species. However, the presence of the four allowed C_{60} vibration bands in the FTIR spectra of solvent treated fullerenes proves that a relatively large amount of C_{60} molecules remain intact during this kind of processing what confirms that the saturation of the interactions process indeed takes place.

III.3. The stability of fullerol water solution in ambient condition

III.3.1. Introduction

The wide spectrum of the fullerene application, ranging from commercial to medicinal, is rapidly expanding. The properly modified nanoparticles in the range of 1–100 nm, including some carbonaceous nanomaterials, such as fullerenes and their derivatives, have been an important topic of research concerning the possible targeting of the cancer cells [103]. However, the extreme hydrophobicity of this sort of molecules being soluble in aromatic solvents but not in water constitutes a serious limitation. Therefore the C_{60} modification with hydrophilic substituents, leading to the increase of their affinity to the aqueous phase, is needed for the future performance in their application [104]. One of the possible C_{60} modifications, drastically increasing its solubility in polar solvents, is its hydroxylation, leading to the formation of the water-soluble class of fullerenes known also as fullerols [104]. This kind of molecules hold a special place among the fullerene derivatives and has been intensively studied taking into account a very wide field of its possible application, ranging from materials for fullerene-containing polymers synthesis, photovoltaic devices to biological and medical research [104–106]. Biocompatible polyhydroxylated fullerenes turned out to be particularly efficient antioxidants acting as free-radical scavengers in several biological systems [103]. The recent studies report moreover that the toxicity of fullerols is lower than that of underivatized colloidal C_{60} aggregates [103,107]. The stability of the derivatized fullerenes in the aqueous phase is one of the crucial factors deciding about its possible application in the field of medicine since the nature of the fullerene–water interaction varies, among others, due to the presence of oxidants and the exposure to light what may take place also in the biological fluids [107]. In this chapter I describe the results of the spectroscopic investigations of the layer obtained from the fullerol water solution stored during 14 months in ambient conditions.

III.3.2. Experimental details
III.3.2.1. Materials and substrates

Fullerol has been obtained from AMD (Dresden). N-type phosphorous doped silicon (100) wafer have been delivered by Crystec, Berlin, Germany.

III.3.2.2. Sample preparation

Fullerol in the concentration of 0.01 % has been dissolved in water. The solution has been mixed for several minutes by means of a mechanical stirrer. The substance obtained in this way has been stored for 14 months in a small bottle made of transparent, colorless glass in ambient conditions. The silicon substrates were cleaned by ultrasonification in acetone, isopropanol and then in distilled water for 10 minutes and were dried afterwards in nitrogen stream. The fullerol solution has been deposited by dropping onto the substrate, afterwards the samples were dried in air. The scheme of the sample is presented in Fig. III.3.1.

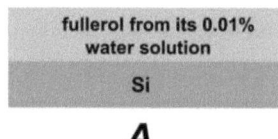

Fig. III.3.1. Scheme of the fullerol sample described in this chapter (referred further in the text as sample *A*).

III.3.3. Results and discussion

III.3.3.1. XPS investigations

The C 1s peak, centred at 284.8 eV, was used to calibrate the spectrum in order to compare the centre positions of the particular components with the literature data [104]. The C 1s core level XPS spectra of the sample A is presented in Fig. III.3.2, whereas the Table III.3.1 lists the binding energies extracted by the peak fitting procedure.

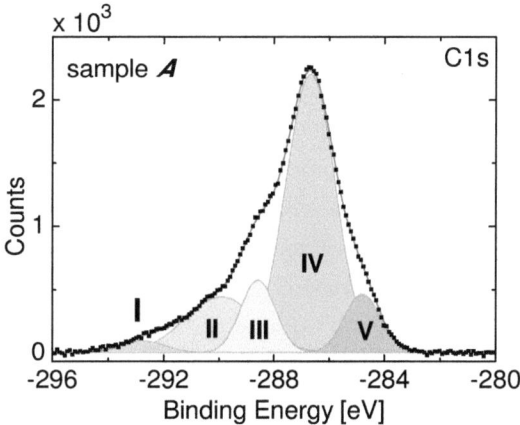

Fig. III.3.2. XPS C1s core level spectrum of the sample A.

Since the surface of the analysed film was not homogeneous, the quantitative XPS analysis has not been performed. The following discussion concerns the qualitative composition of the investigated layer.

Table III.3.1. Center positions found in a consequence of the sample A XPS spectrum decomposition.

peak	centre position, sample A [eV]	literature data	assignments
I	-292.8		shake up satellite[108]
II	-289.9		C=O [105]
III	-288.6	-288.1[104] -288.3 [105]	C–O⁻ [104,105]
IV	-286.7	-286.7[104]	C–OH[104,105]
V	-284.8	-284.8[104]	C–C [104,105]

As shown in Fig. III.3.2 and Table III.3.1, the decomposition of the C 1s core level spectrum of sample *A* allowed the assignment of three different carbon oxidation states (peaks II, III and IV). The source of the peak V is non-oxygenated carbon [105]. Taking into account that the contribution of this feature to the overall area of the C 1s signal is low (around 9.3 %) I conclude, that the majority of the carbon species building the fullerene cages underwent oxygenation. The feature II originating from the hydroxylated carbon represents almost 60 % of the total area of the signal C 1s, giving the hint that the analysed layer contains mostly this kind of species. Although we don't know the preparation technique of the starting material used for sample *A* fabrication, one may suppose, that this fullerol substance was precipitated from a basic aqueous solution, thus the cages were additionally functionalized with -ONa species. Therefore the signal III may be assigned to C–O$^-$ groups [104] or dioxygenated hemiketal carbon [105]. The feature II with the center position around -289.9 eV originates from the carbonyl group [105]. The formation of C=O species has been commonly observed as a result of the oxidation of the fullerene solutions exposed to the light in ambient conditions [80,109]. The small, broad feature I with a center position around -292.8 eV may be assigned to the π-π* transitions [108] within the fullerene cages what would suggest that the analyzed substance still contained a part of not affected fullerene species.

III.3.3.2. NEXAFS investigations

In order to estimate the central position of the most pronounced and prominent resonances, the C-K edge NEXAFS spectra have been decomposed using a possibly low number of the asymmetric P-Voigt functions for the resonance peaks and one arctangent for a step function [83–85]. Fig. III.3.2 illustrates a possible way of the spectra decomposition. Since the spectroscopic assignment is still an object of ongoing exploration and sometimes a quite challenging and controversial discussed task [90], only the most pronounced resonances were considered in order to obtain the information concerning the local bonding.

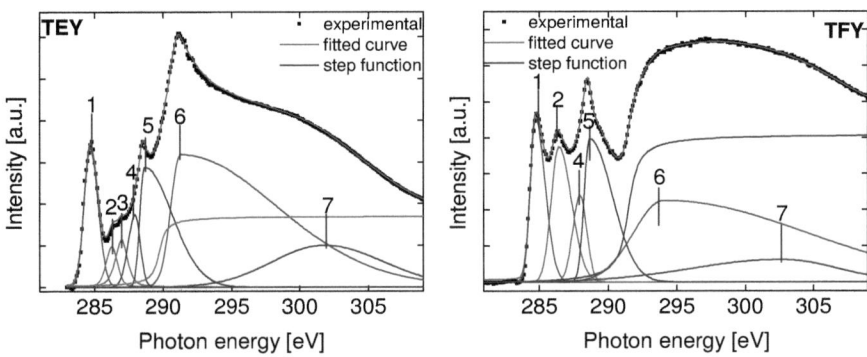

Fig. III.3.2. NEXAFS C K-edge spectra of the sample A recorded in TEY and TFY mode.

Table III.3.2. Center positions of the peaks obtained as a result of the sample A TEY and TFY decomposition.

peak	TEY [eV]	assignment	TFY [eV]	assignment
1	284.7	C1s–π*C=C[90]	284.7	C1s-π*C=C[90]
2	286.2	C1s–π*C=C[110]	286.4	C1s(C–R)-π*C=C[90] C1s(C–OH)-π*C=C[111]
3	287	C1s(C–OH)-π*C=C[111,112]		
4	288	C1s-π*C=O[113]	288	C1s-π*C=O[113]
5	288.6	C1s-π*C=C[110] C1s-π*C=O[90]	288.6	C1s-π*C=C[110] C1s-π*C=O[90]

The K-shell emission resulting in the sharp features observed around 284.7 eV in both TEY and TFY spectra is attributed to the excitations from the C 1s core level into empty states of the π*-electron-system [90]. In general, the presence of these features confirms that a number of the C=C bonds of the fullerene containing system has not been affected while the processing. Since the X-ray fluorescence process is weaker than the non-radiative channels [86], the spectrum recorded in the TEY mode is significantly more pronounced than that obtained in TFY. Therefore the decomposition of TEY allowed to extract two clear resonances at photon energies around 286.2 eV and 287 eV (signals 2 and 3) corresponding to the C 1s-π*C=C [110] and C 1s(C–OH)-π*C=C [111,112] transitions, respectively, whereas in the TFY data one observe only one very prominent feature at the photon energy around 286.4 eV (signal 2) containing most probably the contribution from both transitions mentioned before for the TEY data in that energy range. The signals 4 around 288 eV, observed in the spectra recorded in both modes, are probably induced by the presence of carbonyl groups [113], detected also by XPS. The relatively broad features 5 around 288.6 eV, observable for both discussed spectra, have been assigned to the overlapping C 1s-π*C=C[110] and C 1s-π*C=O[90] transitions. At the higher photon energy one may notice some broader features (6 and 7) corresponding to the C 1s-σ* resonances however they are not prominent enough to clearly specify their center positions, therefore a more detailed discussion regarding their assignment is omitted.

III.3.4. Conclusions

The spectroscopic investigation revealed the presence of different carbon oxidized states within the analyzed sample. XPS exhibits the superiority of the C–OH species proving that the fullerene cages were highly hydroxylated. On the other hand both spectroscopy methods (XPS and NEXAFS) confirmed the presence of carbonyl groups formed most probable due to the storage conditions. Although the majority of the carbon atoms has been affected by the oxygenation, the investigations revealed the existence of a number of C=C species not affected by this process. Taking into account that the analyzed fullerol water solution has been stored more than a year in the ambient condition and moreover has not been protected from the light one may sum up that its stability turned out to be impressively high. The fact that not every carbon species building the fullerene cages have been affected by the interactions with the environment may be explained by the occurrence of the kinetic limitation due to the steric hindrance. This would be at the same time a hint that a large number of the fullerene cages has not been opened and destroyed during the described processing. Another explanation of this kinetic saturation would be the formation of small agglomerates and their further passivation. In this situation the passivation layer would protect the non-reacting fullerol molecules from the environment influence [102]. However, one should take into account that the concentration of the described solution was very low (0.01%) what should not promote the agglomerates formation. In order to investigate this phenomenon in more details the investigation by means of the atomic force microscopy would be desirable.

III.4. Temperature influence on the properties of thin CuPc and C_{60} fullerene films deposited on silicon substrates[2]

III.4.1. Introduction

Amino-functionalization is a common approach in order to engineer silicon dioxide surfaces by the immobilization of inorganic and organic molecules [114]. Due to the self-catalytic nature of the amino group, APTMS relatively easy undergoes hydrolysis even due to the ambient humidity [114], and forms through the further condensation a SiO_2 network with aminopropyl functionalities. This kind of aminoalkoxysilanes is used in such technologically important preparations techniques like ALD in order to functionalize the surface with amino groups [114] or to perform a controlled deposition of the silicon dioxide [115]. However, these processes require a precise temperature control since, according to the literature, above 200 °C significant rearrangements within the films, including the chemical decomposition through the propyl group oxidation and formation of the silicon dioxide at around 300 °C, take place [114,116,117]. Thus, the information concerning the possible temperature range of the surface decomposition is valuable for this deposition technique. The temperature influence is also important in the case of the aminosiloxane based composite materials especially when there is a need to remove the physi- and chemiadsorbed water molecules by heating desired for instance for low-*k* materials [118]. In this chapter I describe the heat treatment influence on the chemical composition of the two following films: thin APTMS layer deposited by spin-coating on a silicon substrate and APTMS based thin composite films, containing CuPc, PCBM and POSS dopants. The organic residual decomposition progress upon annealing in the temperature range of 200 °C-400 °C and its influence on the film quality were monitored by the quantitative analysis of the XPS spectra.

III.4.2. Experimental

III.4.2.1. Materials and substrates

C_{60} (99,5 %), PCBM (99,5 %), APTMS (97 %) and POSS, obtained from Sigma-Aldrich has been used. CuPc (95 %) has been delivered from Alfa Aesar. N-type silicon (100) wafers (Phosphorous doped, Crystec, Berlin, Germany) with a resistivity of 1-5 Ωcm were chosen as substrates for the samples.

---

[2] J. Klocek, K. Henkel, K. Kolanek, E. Zschech, D. Schmeißer, Annealing influence on siloxane based materials incorporated with fullerenes, phthalocyanines and silsesquioxanes, BioNanoScience 2 (2012) 52-58

III.4.2.2. Samples preparation

In order to prepare the films A and B, the Si substrates were cleaned by ultrasonification in isopropanol and then in distilled water for 10 minutes and were dried afterwards in nitrogen stream. The wafer surfaces were hydroxylated by the immersion in piranha solution (a mixture of 7:3 (v/v) 98 % H_2SO_4 and 30 % H_2O_2) at 120 °C for 1.5 h. The substrates were then rinsed with distilled water and isopropanol several times and were dried afterwards in nitrogen stream.

The surface of the sample A consists of pure APTMS. The solution for the film of sample B was obtained by vigorous stirring of the following components: PCBM (0.3 %), CuPc (0.4 %), POSS (0.3 %) and APTMS (99 %) for one hour. No external solvent has been used for the investigated films preparation. The schemes of the samples A and B are drawn in Fig. III.4.1.

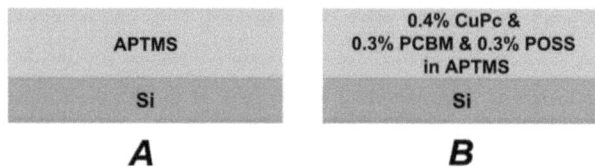

Fig. III.4.1. The schemes of the samples of APTMS based films on silicon substrates: A) pure APTMS (referred in the text as sample A), B) APTMS based composite material incorporated with CuPc, POSS and PCBM molecules (sample B).

In order to estimate the temperature influence on the properties of the prepared films the samples were heated outside of the XPS system in ambient conditions and measured by XPS directly after the heating. Sample A was measured 4 times: without heating (referred further as sample A), after 1 h in 200 °C (sample A_{200}), after 1 h in 250 °C (sample A_{250}) and after 1 h in 300 °C (sample A_{300}). Sample B was measured twice: without annealing (referred further as sample B) and after 1 h in 400 °C (sample B_{400}).

III.4.3. Results and discussion

III.4.3.1. Evolution of the pure APTMS film composition upon heating basing on XPS investigations.

The corrections for the energy shift of the spectra were accomplished by assuming a binding energy of 103.5 eV for the Si^{+4} species within the Si 2p peak [119]. In the first part of this experiment I performed the quantitative XPS analysis of the temperature dependence of the surface chemical composition of pure APTMS. The detailed peak decomposition analysis of the core levels of the sample A at room temperature will be reported in chapter X (referred as sample A therein). Here the deduced atomic composition of the film in dependence on the temperature will be discussed. Fig. III.4.2 shows the evolution of the surface chemical composition of film A as a function of the temperature.

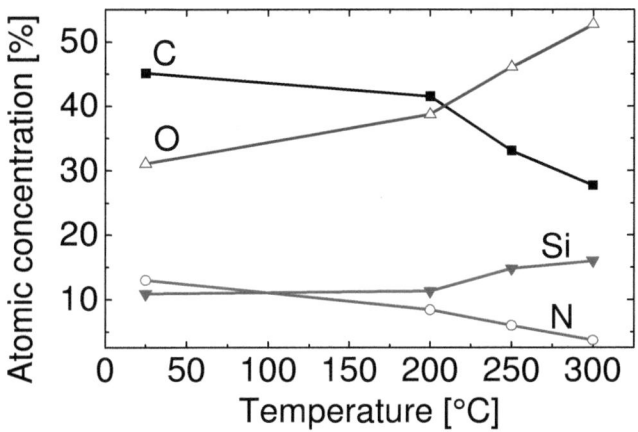

Fig. III.4.2. Temperature dependent chemical composition of the APTMS film surface.

The data plotted in the Fig. III.4.2. indicate clearly that the temperature above 200 °C strongly accelerates the decomposition of organic residuals. The surface carbon atomic concentration measured around 24 h after the preparation and storage in ambient conditions equals 45.1 %, while upon annealing to 200 °C, 250 °C and 300 °C this value decreases to 41.5 %, 33.1 % and 27.7 %, respectively. The decrease of the N species within the surface accompanying the temperature increase is even more pronounced. The XPS quantitative analysis revealed the following N atomic concentrations: 13 %, 8.4 %, 6 % and 3.7 % within the surface E, E_{200}, E_{250} and E_{300} respectively. At this point it seems to be quite interesting to follow the temperature dependent C:N atomic ratio change within the surface. Taking into account, that the –OCH$_3$ groups of the APTMS easily undergo a hydrolysis reaction due to the environmental humidity and are at least partially removed as CH$_3$OH to the ambient during the curing [114,119], it is obvious that the main contribution to the carbon species within the APTMS film is determined by the aminopropyl groups (with the C:N ratio of 3:1). For sample A the C:N ratio measured before the annealing steps equals 3.5:1 reflecting a value closed to pure amino propyl groups. The slight excess of carbon suggests, that after 24 h curing in air not all -OCH$_3$ groups have been removed out of the surface as condensation byproducts. The situation changes drastically during the annealing and the C:N ratio increases to 4.9, 5.5 and 7.6 after the heating at 200 °C, 250 °C and 300 °C, respectively. This indicates the gradual destruction of the amino propyl groups. It seems that at the temperature of 300 °C a significant amount of nitrogen atoms is removed as NH_3 to the ambient leaving oxidized carbon species within the surface. At the same time the intensities of the O 1s and Si 2p peaks clearly increase.

III.4.3.2. Annealing influence on APTMS based hybrid material

In the next step the temperature dependent chemical properties of the APTMS based composite material doped by CuPc, PCBM and POSS have been investigated. A detailed interpretation of the core levels of the sample B at room temperature is given in chapter III.9 (referred as sample C therein). Here the influence of the annealing will be discussed in more detail. Fig. III.4.3 represents the XPS spectra obtained for the sample B that has not been tempered and for sample B_{400} that was annealed at 400°C. The detailed peak analysis is reported in Table III.4.1.

The data presented in Fig. III.4.3 and Table III.4.1 clearly indicate that at the temperature of around 400 °C a complete thermal decomposition of the organic residuals takes place. After the annealing at 400 °C, the signal corresponding to the N 1s core level totally vanished as well as no C 1s features originating from the APTMS organic species (composed of sp^3 hybridized carbon atoms) could be detected. In Table III.4.2 the atomic concentration within the surfaces of the films B and B_{400} are presented.

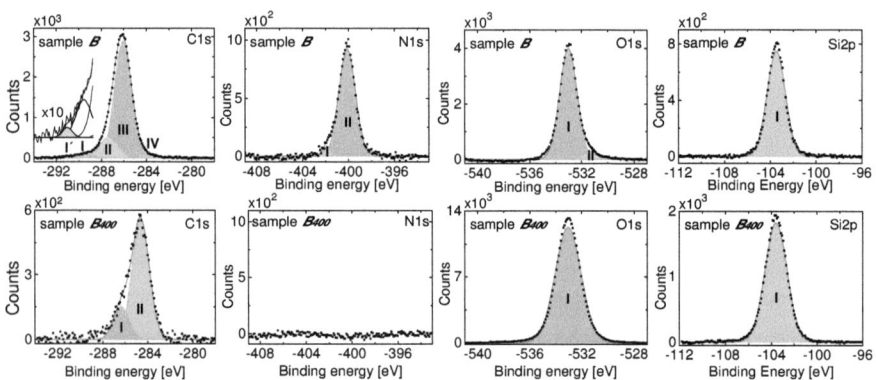

Fig. III.4.3. XPS spectra of the sample B recorded before heating (top row), and B_{400} (bottom row). The following core level spectra are drawn in the same order for every sample: C 1s, N 1s, O 1s, Si 2p. The data of the sample B will also be reported in chapter III. 9.

Taking into account that the basis material of the film B was APTMS one concludes that the thermal rearrangement within that surface is accompanied mainly by the decomposition of the aminopropyl group. The thermal decomposition pathways of this functional group covalently grafted on the silica surface has already been an object of scientific investigations. For this purpose Alekseev and al. [129] applied the methods of temperature programmed desorption mass spectrometry, thermogravimetry and differential thermal analysis. They suggested the possibility of thermal transformations through the interactions between neighboring surface Si–OH and Si–$(CH_2)_3NH_2$ groups resulting in a SiO_2 formation accompanied by ammonia and C_3H_6 releasing during the decomposition reaction.

After the heating at 400 °C the overall C species contribution decreased from around 56.2 % (sample B) to around 7.4 % (sample B_{400}). The C 1s spectrum of the sample B_{400} has been decomposed into two peaks with the center positions at around: -286.3 eV (originating from C–O species [122], signal I) and -284.5 eV (originating from C–C species [63], signal II). The signal I constitutes around 20 % of the overall intensity within the C 1s spectrum while the signal II corresponds to 80 %. The non-oxidized C forms (peak II) originate most likely from fullerene cages and CuPc benzene rings since both species possess high thermal resistances [130,131]. The presence of this signal was not clearly determined for the non-annealed sample due to the significant superiority of the APTMS organic residuals. The absence of any N 1s signal contribution originating from CuPc species in the XPS spectra of the annealed sample is quite puzzling, however it may be attributed to the relatively low concentrations of CuPc within the analyzed surface. On the other hand, Adolphi et al. observed against expectations the destruction of the CuPc molecule structure at a temperature of 350 °C by performing angle-resolved XPS measurements [65]. Basing on our XPS data, this process cannot be neglected also for the sample B_{400}.

Fig. III.4.4 illustrates the scheme of the POSS molecule applied as one of the additional dopants in this part of the experiment.

Table III.4.1. Decomposition of the XPS spectra recorded for the unheated sample B and for the sample B_{400}.

Core level	Peak number		Position		Assignment	
	Sample B	Sample B_{400}	Sample B	B_{400}	Sample B	B_{400}
C 1s	I	-----	-291.0	-----	π-π* shake-up [120]	-----
	I'	-----	-289.6	-----	NHCOO⁻ [119]	-----
	II	-----	-287.4	-----	C–N sp³ [121]&C–O[63]	-----
	-----	I	-----	-286.3	------	C–O [122]
	III	-----	-286.1	-----	–Si–CH₂–CH₂–CH₂–NH₂ [119]	-----
	-----	II	-----	-284.5	------	C–C ; aromatic carbons [123]
	IV	-----	-284	-----	C–Si[124]	-----
N 1s	I	-----	-401.9	-----	NH₃⁺[119]& NHCOO⁻ [119]	-----
	II	-----	-400.1	-----	–NH₂[119]	-----
O 1s	I	I	-533.0	-533.0	Si–O–Si[125]; C–O–C,C–O–O⁻,C–OH[67]	Si–O–Si[125]; C–O–C,C–O–O⁻,C–OH[67]
	II	-----	-531.2	-----	-C=O[126]& -N-C=O[127]	-----
Si 2p	I	I	-103.5	-103.5	O–Si–O/Si–OH[128]	O–Si–O/Si–OH[128]

Table III.4.2. Atomic concentrations within the surfaces of the films B and B_{400}.

Sample	C [%]	N [%]	O [%]	Si⁴⁺ [%]
B	56.2	8.9	24.2	10.7
B_{400}	7.4	0	71.2	21.4

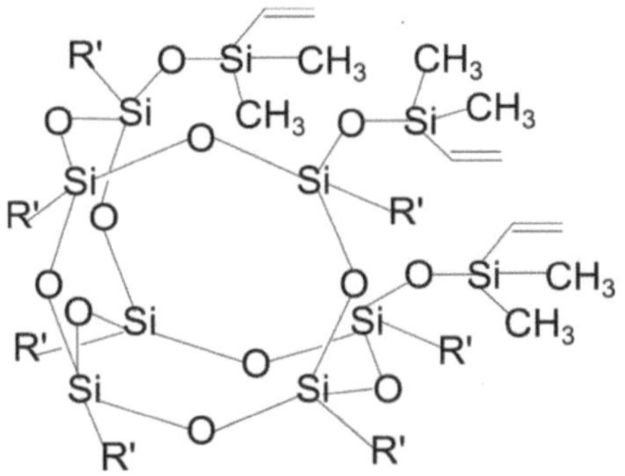

Fig. III.4.4. Scheme of the POSS molecule applied as a dopant to the APTMS matrix.

The thermal decomposition of the cyclopentyl rings, that are organic residuals within the POSS molecules incorporated into the film *B*, occurs around 350-400 °C as reported in [132]. The decomposition of the cyclopentyl species is most probably accompanied by the creation of new Si–O–Si bonds [132]. The POSS molecule applied in the present experiment possess also vinyl functional groups. Alekseev and al. [129] noticed that the thermal transformation of $-C_2H_3$ groups on a SiO_2 surface starts in air above 300 °C and takes place most likely via the addition reaction of oxygen to the vinylic double bonds, while the further annealing leads to the decomposition of the organic structure and to the creation of Si–O–Si bonds. Following the same authors, also the –O–Si–CH_3 groups undergo an air oxidation combined with the releasing of CH_4 groups [129]. In conclusion, the thermal degradation of the organic functional groups of the POSS described above led probably to the formation of –Si–O–Si– bridges between the POSS inorganic cages and the SiO_2 matrix. Thus, one may suppose that the annealing of the film *B* probably caused the entrapping of the POSS inorganic structures into SiO_2 which was formed as a product of the APTMS thermal decomposition.

III.4.4. Conclusions

The experiments revealed that the effect of the temperature is drastic for the chemical composition of the investigated films. One of the crucial conclusions, which may be drawn after the described data analysis concerns the importance of the precise temperature control during the attempts of the surface functionalization with amino groups using APTMS molecules and ALD technique. Te data obtained for the sample *A* clearly exhibits, that above 200 °C this process is already seriously hindered while around 300 °C it becomes almost impossible to be performed due to the temperature degradation of the amino groups.

On the other hand the investigation of the sample B revealed, that the heat treatment of the composite material may be an efficient route in order to obtain a silicon matrix enriched with the sp^2 carbon species. Moreover this kind of heat-treated layers may most probably possess a highly porous structure, formed as a result of the releasing gaseous products of the organic residuals decomposition. Thus, heating the hybrid films containing properly chosen dopants with desired functionalities distributed homogenously within the APTMS monomer may probably lead to a highly porous network enriched with carbon species silica that could find an application for instance as a low-k material.

III.5. Spectroscopic and atomic force microscopy investigations of the hybrid materials composed of the fullerenes and 3-aminopropyl-trimethoxysilane[3]

III.5.1. Introduction

Since their discovery in 1985 fullerenes demonstrated various specific chemical, physical and photo-physical properties [13]. Due to their unique structure, electronic properties, great chemical and thermal stability fullerenes have been considered as a valuable component of the advanced materials [63]. Thin fullerene based films have potential applications in many fundamental technologies related to coating and surface modification ranging from devices for solar energy conversion, biosensing to advanced nanostructured devices for microelectronics [52–54]. However, the limited solubility of the fullerene in the organic solvents imposes the C_{60} modification with other molecules in order to enable its practical application [107]. Since large amount of the covalent fullerene derivatives preserve electronic properties of the parent C_{60} sphere, surface modification with these molecules is of a great interest for the materials science and the semiconductor industry. This may open the possibility of transferring the unique fullerene characteristics to the bulk of the materials [52]. Therefore investigations and understanding of the chemical and physical processes on these systems plays a crucial role in order to realize potential future nanoscale applications by controlled manipulation and modification of interfaces [63].

---

[3] J. Klocek, K. Kolanek, D. Schmeißer, Spectroscopic and atomic force microscopy investigations of hybrid materials composed of fullerenes and 3-aminopropyltrimethoxysilane, Journal of Physics and Chemistry of Solids 73 (2012) 699-706

In this chapter I focused on the investigations of interactions between fullerene C_{60} molecules with 3-aminopropyltrimethoxysilane (APTMS) within the thin films obtained by means of the combined spin-coating and evaporation techniques. I decided to use N containing molecule APTMS in order to increase the affinity of the fullerenes to the siloxane matrix [63]. Chemical reactions of these two components occur mainly due to the covalent attachment of the fullerenes to the terminal amine group from APTMS, following the fact that primary and secondary amines undergo N–H addiction reaction across the C=C bonds in C_{60} which fuse two six-member rings [12]. Since the C_{60} fullerenes have a tendency to form clusters due to van der Waals forces between the particular molecules [107], chemical interactions with the APTMS matrix are supposed to decrease this phenomenon. Thus, from the detailed analysis of the results obtained in described below experiments I tried to estimate among others how coating the substrate with APTMS can influence the formation of thin fullerene layers in the evaporation process. A combination of the spectroscopic and microscopic methods has been used to investigate C_{60} monolayer growth during the evaporation process from both physical and chemical point of view. I measured also the stability of prepared systems in the ambient conditions.

The chapter is organized as follows; basing on the XPS results stability of the thick fullerene films prepared on the APTMS coated substrate is discussed, then the investigations of the interactions between the ultra-thin fullerene layer and continuous APTMS film are described. In the next section, NEXAFS results obtained for the samples composed of the C_{60} and APTMS is discussed and compared to the similar data of the pure fullerene layer. The last experimental part describes NC-AFM surface morphology measurements of the pure APTMS film, ultra-thin C_{60} layer on APTMS and thick fullerene layer on the APTMS. Finally, a summary of the relevant conclusions will close the chapter.

III.5.2. Experimental

III.5.2.1. Materials and substrates

In this work I used C_{60} (99,5 %) and APTMS (97 %) obtained from Sigma-Aldrich. Two types of substrates: Mo and p-type Si(001) were employed. As the APTMS builds a silica matrix in the initial phase of the experiments the material under study was deposited on the Mo surface in order to estimate feasibility of the coating process. It has been assumed that an evidence of APTMS deposition on Si(001) substrate may be confused with presence of the SiO_2 native oxide when XPS technique is utilized. The preliminary experiments on Mo surface proved, that the film was coated homogenously on the whole substrate area and was thick. The quality of the samples obtained on the Mo surface turned out to be high enough to perform XPS and NEXAFS investigations. In order to achieve the further progress of the experiment a more appropriate Si(001) surface has been applied.

III.5.2.2. Sample preparation

In Fig. III.5.1 the schemes of the all investigated samples in this chapter are presented.

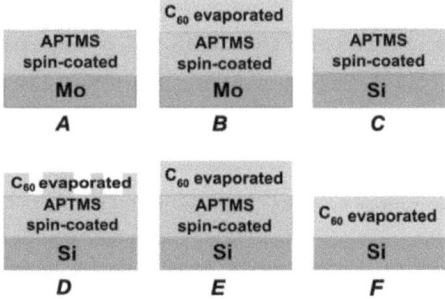

Fig. III.5.1. The schemes of the investigated samples: A) APTMS spin-coated on the Mo surface (referred in the text as sample *A*), B) thick C_{60} layer evaporated on the Mo surface spin-coated with the APTMS (sample *B*), C) APTMS spin-coated on the Si surface (sample *C*), D) ultra-thin C_{60} layer evaporated on Si |APTMS substrate (sample *D*), E) thick C_{60} layer evaporated on the APTMS surface (sample *E*), F) thick C_{60} layer evaporated on the silicon surface (sample *F*).

After the ultrasonification in isopronanol and acetone for 5 minutes, substrates were dried in the nitrogen stream. APTMS was deposited by means of the spin-coating technique at a speed of 6000 rpm for 1 minute in the ambient conditions. The fullerene molecules were deposited on the APTMS coated substrates by evaporation at a temperature of 800°C and a pressure of 1×10^{-5} mbar. In the experiments the evaporation time was varied in order to obtain different thicknesses of C_{60} layers ranging from the ultra-thin layer fabricated by 2 minutes of evaporation (sample *D*), to the thick layers obtained by 10 minutes of evaporation (samples *B* and *E*). The reference sample *F* was produced by the fullerene powder evaporation onto a Si(001) surface at the same conditions as before but for 120 minutes in order to obtain a thick film.

III.5.3. Results and discussion

III.5.3.1. XPS measurements

Since XPS allows to evaluate the chemical composition of the surface [25] this tool has been applied in order to study the C and N chemical states. Additionally I also performed the analysis of the O and Si core levels. For all measured films quantitative XPS analysis has been done. Corrections for the energy shift were accomplished by assuming 103.5 eV binding energy for the Si 2p peak of the Si^{+4} species [119].

III.5.3.1.1. Samples deposited on the Mo surface

In this part of the experiment sample A [Fig. III.5.2(a-d)] and B [Fig.III.5.2(e-h)] were investigated. Sample B was investigated twice: first time in situ direct after the C_{60} deposition (Fig. III.5.2h) and second time after 9 months of the storage in the air [Fig. III.5.2(e-g)]. The main attempt of this measurements was to investigate the ambient influence on the thick fullerene film deposited on the Mo substrate coated with APTMS.

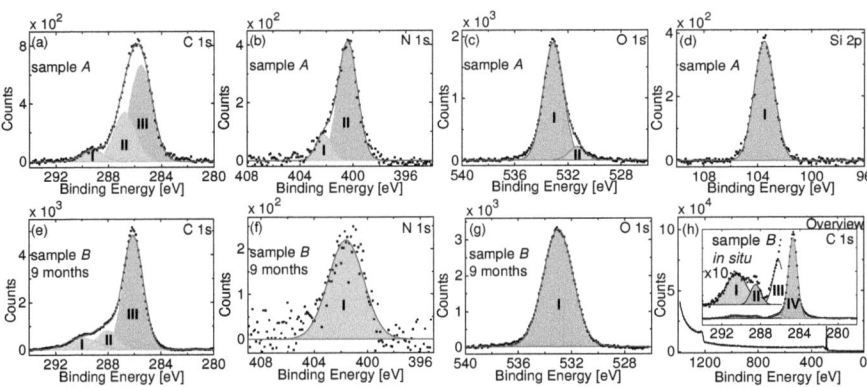

Fig. III.5.2. XPS spectra of the samples prepared on the Mo surface. Top row (a-d) represents the sample A, bottom row (e-g) depicts the sample B after 9 months of storage in the air, (h) shows wide scan survey of in situ prepared sample B and inset depicts C 1s core level (Al Kα 1486.6 eV X-Ray source was utilized).

After the XPS spectra evaluation, the quantitative analysis of the investigated films was performed. The spectra normalization was achieved by dividing the observed relative peak areas by atomic and instrument sensitivity factors equal to 0.711, 0.296, 0.477, and 0.339 for O, C, N, and Si, respectively [25]. After the analysis the following composition in atomic percent has been obtained: C=45.2 %, N=10.8 %, O=32.1 %, Si=11.9 % for the sample A and C=73.6 %, N=2.8 %, O=23.7 % for the sample B after the storage in the air. The stoichiometry of an unhydrolyzed APTMS monomer (excluding H atoms) corresponds to C_6NSiO_3, in the case of the sample A the calculated stoichiometry was around $C_{3.8}N_{0.9}SiO_{2.7}$. Taking into account that the "ideal" poly[(aminopropyl)siloxane] polymer has a stoichiometry of $C_3NSiO_{1.5}$ [119] one may conclude, that in the case of the sample A the monomers have partially undergone the condensation reactions during the spin-coating and drying process.

Since XPS provides information about the chemical states of the elements within the analysed surface [70], the qualitative analysis basing on the particular peak shifts has been performed. Results of the analysis are reported in Table III.5.1.

Table III.5.1. XPS investigations of the samples prepared on the Mo surface: sample A measured after the preparation and sample B after 9 months of the storage in the air.

Region	Peak number		Peak position [eV]		Percentage of the main peak area [%]		Assignment
	Sample A	Sample B	Sample A	Sample B	Sample A	Sample B	
C 1s	I	I	-289.2	-289.2	7.6	13.3	$-NHCOO^-$ [119]
	II	II	-287.2	-287.9	31.1	7.1	C–N sp^3[121] C–O[63]
	III	III	-285.7	-286.1	61.1	79.6	C–C, C–H
N 1s	I	I	-402.3	-401.6	14.9	100	$-NH_3^+$[119]
	II	---	-400.4	---	85.1	---	$-NH_2$[63]
O 1s	I	I	-533.1	-533	91.4	100	–C–O[127], –Si–O[128]
	II	---	-531.3	---	8.6	---	–N–C=O[127]
Si 2p	I	I	-103.5	---	100	---	O–Si–O/ Si–OH[128]

In the sample A the N 1s core level peak located at around -402.3 eV binding energy (signal I) is attributed to the positively charged quaternary nitrogen of the group $-NH_3^+$ [133][134]. The presence of this species are explained by the following reactions [119]:

$$\equiv Si(CH_2)_3NH_2 + CO_2 \leftrightarrow \, \equiv Si(CH_2)_3NH_2^+COO^-$$

(III.5.1)

$$\equiv Si(CH_2)_3^-NH_2^+COO^- + NH_2(CH_2)_3Si \equiv \, \leftrightarrow \, \equiv Si(CH_2)_3NHCOO^- + NH_3^+(CH_2)_3Si \equiv$$

(III.5.2)

The presence of the peak at -289.2 eV in the C 1s core level (signal I) confirms the (III.5.1) and (III.5.2) reactions since it indicates the existence of the carbamate species that forms as a product of the interactions between the amine group from APTMS with CO_2 at the film-air interface [11].

Sample B was prepared by evaporation of thick C_{60} layer (thickness more than 10 nm) on the Mo |APTMS substrate, thus it's in situ XPS investigations (i.e. directly after deposition), exhibit only C species (Fig. III. 5.2h). Taking into account short evaporation time (of 10 minutes) one may conclude that APTMS significantly accelerates the process of thick fullerene layer formation on the surface. In order to obtain similar thickness of the C_{60} layer on the pure Si(001) surface (sample F) the evaporation process had to be carried out for at least 60 min (XPS spectrum not shown). The binding energy of the signals I, II, III, IV in the C 1s core level around: -290.5 eV, -288.5 eV, -286.2 eV and -284.7 eV (Fig. III. 5. 2h) corresponds to the carbon π-π* shakeup peak, C=O species from the surface, C–O bond coming probably from the partial oxidation and C–C bonds of the fullerene molecules respectively [63].

An interesting phenomenon is observed on the sample B after the long-time storage in the air [Fig. III.5.2(e-g)]. On the surface not only C features were identified, but also pronounced O peak and weak N signal are present. The calculated stoichiometry of the surface is $CN_{0.1}O_{0.3}$. Since the solid fullerenes are air-sensitive materials [135], the presence of the O feature in the overall XPS spectrum confirms the oxidation reaction between C_{60} and the oxygen from the air. Nevertheless, rather unexpected is the presence of the N atoms in the near surface region, while at the same time no Si features are detected (not shown). I explain this phenomenon by the diffusion of the volatile $-NH_3$ groups, formed as a result of APTMS decomposition, towards the surface. More details regarding these decomposition pathways are described in [119]. I presume, that these nitrogen containing groups have been chemisorbed and/or physisorbed within the fullerene film. Hence, originating from APTMS volatile nitrogen groups have been attached to the surface through the interactions with the fullerene molecules.

III.5.3.1.2. Samples deposited on the Si(001) surface

I investigated sample C [Fig. III.5.3(a-d)] and D which was measured in situ direct after the deposition [Fig. III.5.3(e-h)]. Detailed peak analysis is shown in Table III.5.2.

The quantitative analysis reveals following surface composition of the samples in atomic percent: C=46.9 %, N=10.4 %, O=32.6 %, Si=10.1 % for the sample C and C=52 %, N=9 %, O=29 % and Si=9.8 % for the sample D. Pure APTMS film of the sample C contains slightly higher amount of the C and O species in comparison to the similar sample prepared on the Mo surface (sample A) what indicates that on Si(001) the condensation process within

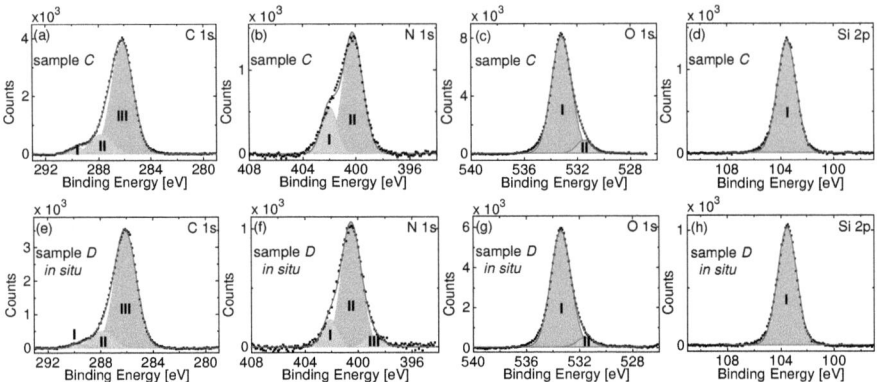

Fig. III.5.3. XPS spectra of the samples prepared on the Si(001) surface. (a-d) Represents sample *C*, (e-h) depicts the sample *D*.

the material was less pronounced. We noticed that the sample *D* contains more C species than the sample *C* what confirms successful evaporation of the C_{60}.

Table III.5.2. XPS investigations of the samples *C* and *D* prepared on the Si(001) surface.

Region	Peak number		Peak position [eV]		Percentage of the main peak area [%]		Assignment
	Sample *C*	Sample *D*	Sample *C*	Sample *D*	Sample *C*	Sample *D*	
C 1s	I	I	-289.3	-289.8	3	1.8	–NHCOO⁻[119]
	II	II	-287.6	-287.9	14.6	12.6	C–N sp^3[121], C–O[63]
	III	III	-286.1	-286.1	82.4	85.6	C–C, C–H[119]
N 1s	I	I	-401.9	402.2	24.3	14.3	$-NH_3^+$[119]
	II	II	-400.2	-400.5	75.7	79.4	$-NH_2$[63]
	---	III	---	-398	---	6.3	$-NH-C_{60}$[63]
O 1s	I	I	-533.2	-533.4	92.2	94	O–H [13]
	II	II	-531.5	-531.6	7.8	6	C–O[14] Si–O[15]
Si 2p	I	I	-103.5	-103.5	100	100	Si–O[17]

Taking into account that the quantity of the evaporated fullerene molecules in the case of the sample *D* was small and the overall content of the carbon species within the film equals 52 % it is difficult to distinguish the features corresponding to the C_{60} molecules in the C 1s XPS spectrum.

Nevertheless, while performing detailed analysis of the spectrum in the N 1s region one can notice evidence of the existing small amount of the fullerene molecules bonded to the amino groups (Fig. III.5.3f). After deconvolution of the N 1s signal one obtain three features in the binding energy range of: (I) -402.2 eV, (II) -400.5 eV, and (III) -398.0 eV. Signals I and II correspond to the positively charged amino group and free primary amino group respectively as it was in the case of the sample C. Additional signal III with a very small intensity is assigned to the primary amino groups bonded to the fullerene molecules. The presence of small amount of C_{60} molecules was also confirmed by AFM analysis (described below).

III.5.3.2. NEXAFS measurements

For the NEXAFS measurements samples: *A*, *B*, and *F* were used. All samples were measured after 4 months of the exposure to the air. Position, width and intensity of the π^* and σ^* resonances in the NEXAFS spectra give information of the changes in the electronic state of the investigated material [28]. Moreover, NEXAFS spectroscopy allows obtaining characteristic features that correspond to sp^3 and sp^2 hybridizations of carbon-based materials [136]. For that reason performing NEXAFS measurements seems to be very helpful in order to investigate the electronic structure of the carbon materials.

Deconvolution of the NEXAFS spectra (not shown here) was performed in WinXAFS software using asymmetric P-Voigt functions for the resonance peaks and an arctangent for a step function [83]. First I focus on the peak assignment of the NEXAFS resonances and the comparison with literature data. In Fig. III.5.4 are shown the results of the NEXAFS measurements together with identified peak positions. Detailed peaks assignments in C 1s edge for spectra obtained in TEY and TFY mode are given in Table III.5.3a and III.5.3b, respectively. I focus on the features appearing below 290.5 eV photon energy that enclose a π^* character in C_{60} molecule [136].

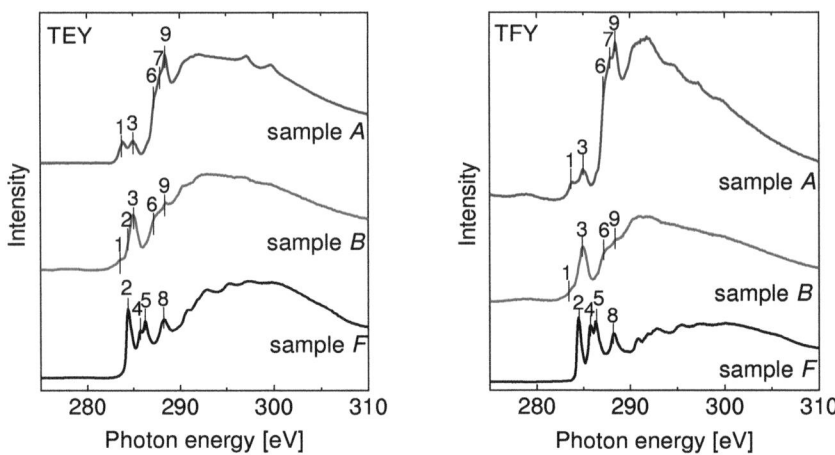

Fig. III.5.4. NEXAFS spectra comparison of: sample A (blue line), sample B (red line) and reference sample F (black line).

Table III.5.3a. Peak assignments for samples A, B, and F measured by means of NEXAFS in TEY mode.

Signal	Energy [eV]			Assigment
	sample A	sample B	sample F	
1	283.9	283.8	----------	C 1s–π*[137]
2	----------	284.3	284.3	C 1s-$\pi^*_{C=C}$ [138]
3	285	285	----------	C 1s(C–H)-$\pi^*_{C=C}$ [90]
4	----------	----------	285.7	C 1s-$\pi^*_{C=C}$ [88]
5	----------	----------	286.2	C 1s-$\pi^*_{C=C}$ [110]
6	287	286.9	----------	C 1s-σ^*_{C-H} [92]
7	287.9	----------	----------	C 1s-σ^*_{C-H} [139] [140]
8	----------	----------	288.1	C 1s-$\pi^*_{R-(C=O)-R}$ [141]
9	288.5	288.4	----------	C 1s (COO$^-$)-$\pi^*_{C=C}$ [142]

54

Table III.5.3b. Peak assignments for samples A, B, and F measured by means of NEXAFS in TFY mode.

Signal	Energy [eV]			Assigment
	sample A	sample B	sample F	
1	283.9	283.9	------------	C 1s-π*[137]
2	------------	------------	284.3	C 1s-π*$_{C=C}$ [138]
3	285	285	------------	C 1s(C–H)-π*$_{C=C}$ [90]
4	------------	------------	285.7	C 1s-π*$_{C=C}$[88]
5	------------	------------	286.2	C 1s-π*$_{C=C}$ [110]
6	287.2	287.1	------------	C 1s-σ*$_{C-H}$ [92]
7	288	------------	------------	C 1s-σ*$_{C-H}$ [139,140]
8	------------	------------	288.2	C 1s-π* [141,143] C 1s-π*$_{R-(C=O)-R}$ [93] C 1s-π*$_{(C=O)}$ [142]
9	288.5	288.5	------------	C 1s (COO$^-$)-π* $_{C=C}$ [141]

The peaks at around 283.8 eV (signal 1) measured in both TEY and TFY modes recorded for the sample A and B are identified as carbon atoms with the sp^2 hybridization. The same features were also found in the NEXAFS spectra of natural diamond [137,144]. The broadening of the peak number 2 in TEY spectra of the sample F is probably also caused by the sp^2 contaminations.

The reference C$_{60}$ sample F exhibits intensive resonances at around 284.3 eV (signals 2 for both TEY and TFY mode). These values correspond to the C 1s lowest unoccupied molecular orbital resonance energy for fullerenes [145]. Two another features that can be noticed only for the sample F and are attributed to the C 1s-π*$_{C=C}$ transitions occurring at around 285.7 and 286.2 eV (signals 4 and 5, respectively). While analysing those features one may clearly notice, that the signal 4 is more pronounced for the spectrum recorded in the bulk sensitive TFY mode. This phenomenon is quite easy to explain since the surface of the sample F was more affected by the oxidation during the storage in the ambient conditions than the bulk. As a result of oxidation some of C=C bonds break and the intensity of the TEY signal 4 decreases.

Features corresponding to the resonance at around 284.3 eV are probably the reason for the significant broadening of the main peaks at around 285 eV for the sample B. The low intensity of the π* resonances in the NEXAFS spectra of the sample B might be explained by the chemical reaction of the fullerenes with O and N atoms from APTMS that reduce the number of the π bonds within the material and at the same time leads to the fewer π* transitions. Features at around 285 eV indicate C 1s (C–H)-π*$_{C=C}$ transitions [90] and appear in the spectra of samples A and B recorded in both TEY and TFY mode (signals 3). As in the case of the fullerene based sample *B* the existence of the π* transitions might be attributed to the presence of the π bonds within the C_{60}, those features are quite surprising in the case of the sample *A* since the APTMS nominally forms a saturated system. However this phenomenon was also observed by Graf et. al in the NEXAFS spectra of the 3-aminopropyl-triethoxysilane film and they have attributed it to the unsaturated species formed due to the radiation damage during NEXAFS measurement [139]. Additionally, partial contamination of the substrates used for the sample's preparation was not excluded.

Resonances covering the range 287.1-287.9 eV might also provide some valuable information concerning the structure of the investigated material and usually correspond to the C 1s (C–H)-σ*$_{C-H}$ transitions within saturated, organic, aliphatic molecules [139]. Features located in this range are observed for the sample *A* (signals 6 and 7) and for the sample *B* (signals 6). Broad shape of the signal 6 in the spectra of the sample *B* and the shift of its centre positions of around 0.1 eV to the lower photon energy in comparison to the sample *A* might suggest that they contain also additional contribution of the π* resonances. According to the literature C 1s (C–R)-π*$_{C=C}$ transitions (where the R is a functional group) evidenced around 286.9 eV are typical for the amine groups [146]. One may presume that in the case of the sample *B* signal 6 of the NEXAFS spectra include the contribution of σ* transitions from APTMS and π* resonances from the amine functionalized fullerenes.

Pronounced feature 8 that can be clearly noticed only for the sample *F* occurs around 288.1 eV and 288.2 eV for the spectra recorded in TEY and TFY mode, respectively. NEXAFS features at 288.2 eV are characteristic of π* resonances [143] and might be attributed to the C 1s-π*$_{R-(C=O)-R}$ transition [93] within the partially oxygenated fullerene molecule. The broadening of this feature in the spectrum recorded for the TEY mode and its shift of 0.1 eV to the lower photon energy might suggest that it consist of the two main states: C 1s-π*$_{R-(C=O)-R}$ excitations within the fullerene molecules and C 1s-σ*$_{C-H}$ /Rydberg transitions (normally observed at 288 eV) [140] attributed to the surface hydrocarbon contaminations. Resonances at around 288.4-288.5 eV (signal 9) existing for samples *A* and *B* are due to the C 1s (COO⁻)-π* transitions [141,142] and confirm that reactions (1) and (2) indeed take place within the APTMS containing films.

NEXAFS spectra for the sample *B* at the C K-edge reveals definitely smaller amount of the π*$_{C=C}$ transitions than the sample *F*, what is evidence of direct interactions between C_{60} and APTMS. This behavior is also in agreement with the XPS results, where for the same sample after the storage in the ambient conditions the C 1s signal shifted to the direction of the higher binding energy in comparison to the spectra of the same sample obtained in situ directly after the preparation process.

III.5.3.3. NC-AFM measurements

Following the detailed chemical analysis of the deposited films NC-AFM has been applied in order to study the surface morphology of the samples: C, D, E (Fig. III.5.1). The main idea was to investigate changes of the APTMS surface texture after the C_{60} evaporation. Measurements were performed on a Si(001) substrate with the native oxide covered by a thin film of investigated materials (Fig. III.5.5). Samples were prepared as described in experimental part. At least three images in different positions on the samples were acquired and fluctuations of the root mean square (RMS) surface roughness (S_q) values of less than ± 7 % were observed. The starting APTMS surface morphology is shown in Fig. III.5.5a (sample D). The calculated S_q equal to 0.20 nm is comparable with result reported in [147] and is similar to a roughness of Si(001) surface covered with native oxide [148,149]. Additional useful information about the surface morphology may be obtained from investigation of the surface height histograms. Instead of specify the whole shape of the height histogram curve it is more convenient to break it down using the moments of the curve. For surface texture analysis of high importance are third and fourth central moments labelled skewness and kurtosis, respectively [150].

Principally the surface skewness (S_{sk}) indicates the degree of symmetry of the height histogram curve while the surface kurtosis (S_{ku}) describes the sharpness of the curve [151]. For Gaussian surface S_{sk} and S_{ku} parameters are equal to 0 and 3, respectively. In the experiment we observed that APTMS after deposition forms a near-Gaussian surface with surface skewness and kurtosis equal to 0.02 and 3.14, respectively. The presence of near-Gaussian surface confirmed by fitting of the surface height histogram with Gaussian curve (Fig. III. 5. 5b) indicates random and homogenous deposition process. Height standard deviation of the Gaussian curve (σ) was estimated to 0.20 nm and is identical to S_q value. In the next step we evaporated the C_{60} fullerenes for 2 minutes at 800 °C (Fig. III.5.5c, sample D). The surface morphology changed notably after the evaporation and a formation of C_{60} clusters with width varied from about 50 nm to 200 nm was observed. S_q increased to 0.34 nm and surface changed its state to less Gaussian with S_{sk} and S_{ku} parameters equal to 0.69 and 4.82, respectively. In particular, the positive skewness indicates the asymmetry of the height histogram and is a clear evidence of the C_{60} clusters on the surface. The tendency of C_{60} clusters build-up was reported in [63,107]. Surface height histogram presented in Fig. III.5.5d clearly reveals a bimodal distribution and the two features marked with light red and green peaks are attributed to APTMS and C_{60}, respectively. Surface height histogram is a very accurate tool for calculation of the height of the features present on the substrate [152,153]. The height (*d*) of the C_{60} clusters equal to 0.65 nm was calculated as a difference between APTMS peak and C_{60} peak centre positions and is in very good agreement with the measured C_{60} diameter [154,155]. This proofs that for short time of evaporation clusters of C_{60} are arranged in islands of one monolayer thickness. Analysis of the peaks area ratio may deliver information about the C_{60} coverage [156]. As a result of the calculations C_{60} coverage was estimated to about 5.1 %. For the bimodal distribution it is also possible to obtain the σ values separately for both materials [157]. These values are equal to 0.25 nm and 0.30 nm for APTMS (σ_r) and C_{60} (σ_g), respectively. Summarizing the results we may conclude that the surface morphology indicates not closed layer of C_{60} on APTMS. In contrast, after 10 minutes of C_{60} evaporation the surface morphology changed significantly and randomly distributed clusters (S_{sk} = 0.07, S_{ku} = 3.00) of a mean height of 7 nm, width of 35 nm, and a period of 100 nm were observed (Fig. III.5.5e, sample *E*). Surface roughness was equal to 3.49 nm and σ was estimated to about 3.57 nm basing on the surface height histogram analysis (Fig. III.5.5f). The surface texture indicates a closed layer of C_{60} molecules on APTMS surface. The observed C_{60} granular morphology is in agreement with results reported in [158,159].

III.5.4. Conclusions

Spectroscopic investigations of the fullerene films deposited by evaporation on the APTMS coated substrate indicate a strong chemical interactions between electrophilic C_{60} molecules and the nucleophilic amine groups. Significant decrease of the π^* resonances in the NEXAFS spectra of the fullerene evaporated on the APTMS film confirms that the majority of the double bonds within the fullerene molecules were broken due to the chemical reactions.

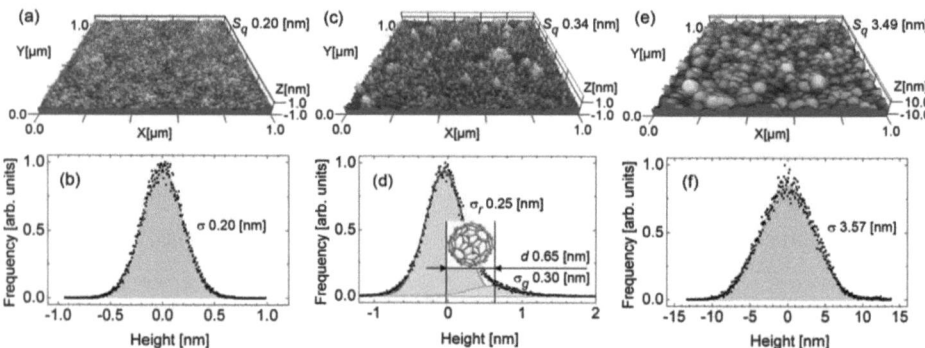

Fig. III.5.5. NC-AFM investigations of the surface morphology and height histograms of the: (a, b) sample *C*, (c, d) sample *D*, (e, f) sample *E*. RMS surface roughness S_q, height standard deviations σ and average C_{60} cluster height *d* are also reported. The raw height data is levelled using the plane correction in order to set the mean value of the image to zero. The height histograms curves are normalized to the same maximum level. The colour scale equals 2 nm, 4 nm, and 27 nm, for samples *C*, *D*, and *E,* respectively.

Information provided by both XPS and NEXAFS spectroscopy complement each other and suggest the presence of the amino functionalised fullerene species formed as a result of interactions between spin-coated APTMS film and the evaporated C_{60}. This is in the agreement with the following chemical reaction described by Bell et al. [15]:

$$C_{60} + nH_2NCH_2CH_2CH_2Si(OMe)_3 \rightarrow C_{60}Hn(NCH_2CH_2CH_2Si(OMe)_3)_n$$

$(III.5.3)$

Due to the high affinity of the fullerene molecules to the amino groups, layer of the APTMS molecules on the substrate accelerates the formation of the C_{60} film during the evaporation. AFM turned out to be a great support for the spectroscopy technique in order to investigate this process in a more details and allowed to detect the formation of a particular islands of the C_{60} monolayers after short time of fullerene evaporation. The measured island thickness reveals that fullerene diameter is not affected by the interactions with APTMS. Longer evaporation time produces thick fullerene layer that organises in a large C_{60} clusters. AFM results exhibit, that RMS surface roughness of the deposited APTMS is similar to the roughness of SiO_2 native oxide.

The investigations of the fullerene based sample after the storage in the air suggest that due to the air exposure many chemical reactions take place within the hybrid material. The results of the XPS investigations shows clearly that this kind of the samples containing both fullerene and APTMS components are not resistant to the air and undergo a strong oxidation.

III.6. Fullerene based materials for ultra-low-*k* application obtained by the means of the sol-gel method[4]

III.6.1. Introduction

The rapid development of microelectronic performance follows the Moore's law predicting that the number of transistors in the integrated circuits increases twice every eighteen months [7].

---

[4] K. Broczkowska, J. Klocek, D. Friedrich, K. Henkel, K. Kolanek, A. Urbanowicz, D. Schmeißer, M. Miller, E. Zschech, Fullerene based materials for ultra-low-k application, Students and Young Scientists Workshop "Photonics and Microsystems", 2010 International. S. 39-43. DOI: 10.1109/STYSW.2010.5714165.

However, because of the huge transistor's density in the advanced integrated circuits its size has to be reduced what makes the circuits faster while at the same time the overall resistance-capacitance (RC) delay increases. One of the strategies of increasing the speed of microelectronics' devices is to reduce the RC delay [8,9]. One of the strategies to reduce RC delay is introduction of low-k materials as an isolation between the nano-wires in microelectronic devices. From a theoretical point of view, fullerene materials are very promising candidates for low-k dielectric materials. The results of calculations show that cross-linking of C_{60} fullerenes by suitable bridge molecules is an efficient way to construct materials with ultra-low dielectric constant [61]. Fullerenes, since theirs' discovery in 1985, demonstrated a various applications on many fields of science, however there are few reasons such as: ability to form clusters, low solubility and thermal stability, that imposes theirs' modification with other materials in order to enable the practical application [160,107]. In this chapter results of spectroscopic investigations of sample produced in Wroclaw University of Technology by incorporation of C_{60} molecules into the material produced by the means of the sol-gel method have been reported.

III.6.2. Experimental

III.6.2.1. Materials

Phenyl triethoxysilane (PhTriEOS) from Aldrich (98 % purity), methyl trimethoxysilane (MeTriMOS) from Fluka (purum) were used. Ethanol (EtOH), acetone and toluene were obtained from POCh (Poland) and used as received. Hydrochloric acid (HCl), available from POCh was diluted in distilled water before use. C_{60} fullerenes (99 % purity) were obtained from MER Corporation (Arizona, USA) and used as received as well.

III.6.2.2. Sample preparation

A basic sol solution was prepared by mixing PhTriEOS : MeTriMOS : EtOH : HCl in molar ratio 1.00 : 1.22 : 7.77 : 1.03 × 10^{-3} respectively. Then a saturated solution of C_{60} in toluene was prepared with concentration 2.80 mg/ml. C_{60} solution was mixed with the basic solution PhTriEOS : C_{60} in molar ratio 1.00 : 3.50 × 10^{-3}, respectively. Afterwards the substrate was coated from such a solution and dried. The detailed procedure, previously used for preparation of similar thin film obtained from different precursors, is described in literature [161]. As a substrate a monocrystal silicon wafer with a 100 nm layer of SiO_2 was used. The substrate was rinsed in acetone, distilled water and finally in ethanol, then dried at room temperature before coating.

III.6.3. Results and discussion

III.6.3.1. XPS measurements

Chemical composition of obtained film was evaluated by XPS analysis. The XPS spectra of regions C 1s, O 1s and Si 2p are given in Fig. III.6.1. On the basis of the Gaussian functions spectra were deconvoluted into more peaks that correspond to different binding energies. The fitting results of the obtained spectra and their assignments are given in Table III.6.1.

After the normalization of XPS spectra, the ratio of the elements contained in the investigated film was estimated. Spectra normalization were performed by dividing their areas by atomic sensitivity factors equal to 0.711, 0.296, and 0.283 for O1s, C1s, and Si2p, respectively. Basing on XPS analysis the ratio of particular elements within prepared film C:O:Si equals 2.7:1.9:1. According to those results, the ratio of silicon to oxygen atoms within examined material is very similar to SiO_2 and equals 1:1.9. Formation of the SiO_2 network is typical for the materials obtained by the sol-gel method. However, in the case of investigated sample the existence of the carbon atoms is of high importance. The contributions of the carbon in XPS spectra come not only from fullerene molecules incorporated into the material but mainly from aromatic rings in Ph-TriEOS and from methyl groups from Me-TriMOS and Ph-TriEOS that were not removed during the fabrication process.

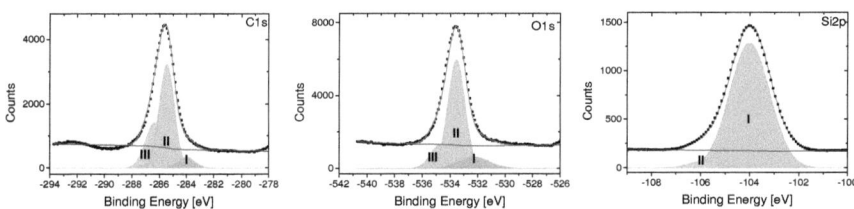

Fig. III.6.1. XPS spectra of fullerene based material produced by sol-gel method. Regions: C 1s, O 1s, and Si 2p are shown.

Table III.6.1. XPS investigations of the sample of fullerene based material produced by sol-gel method. Peak position were identified according to the literature cited in the table.

Region	Peak number	Position [eV]	Percentage of the main peak area [%]	Assignment
C 1s	I	-284	7.12	C–C sp^2 [162]
	II	-285.46	61.41	C–C sp^3 [70], C–C/C–H [72,163]
	III	-286.36	31.47	C–O [141,163]
O 1s	I	-532.23	12.68	O–H [164]
	II	-533.55	70.96	C–O[127], Si–O[128]
	III	-534.74	16.37	CO$_2$[165]
Si 2p	I	-104	100	Si–O[166]

III.6.3.2. NEXAFS measurements

NEXAFS data of the fullerene based sample are displayed in Fig. III.6.2. (solid line) in comparison with a pure C_{60} film (dash line). Here, the TFY delivers bulk information while TEY is more surface sensitive.

In the case of C_{60} derived components, there is a particular mechanism causing a characteristic feature in the absorption data. Based on the electronic properties of the C_{60} molecules in all C_{60} derived compounds, there is a characteristic split in the lowest absorption band causing a three-peak structure [47,48]. In analyzed sample three peaks transitions located at 284.9 eV, 287.5 eV and 288.5 eV are found. In particular, the middle band appears at a higher photon energy than in pure C_{60} films.

Considering small concentration of the fullerenes within the sol-gel material and significant shift of the middle band one may presume that the main contribution to the features corresponding to the C1s-π* transitions comes from the phenyl groups of one of the precursors (Ph-TriMOS). According to the literature features in the range of 284 to 289 eV may be assigned to the C1s-π* transitions of the benzene derivatives [142]. Broadening of the feature near 287 eV may be caused by the contribution of the C 1s-σ* (C–H) resonances from the methyl group [92]. Nevertheless, one cannot exclude the contribution of the C1s-π* transitions of the fullerenes in the NEXAFS spectra as shown in Fig. III.6.2.

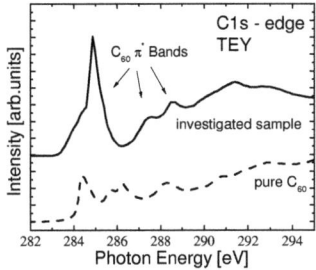 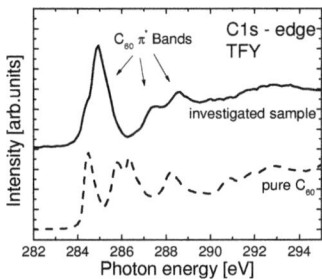

Fig. III.6.2. Comparison of NEXAFS data of samples fabricated by evaporation (dashed line) and sol-gel method (solid line), respectively: the spectra show the typical C_{60} absorption features on the surface TEY and also in the bulk TFY.

III.6.3.3. FTIR measurements

Fourier transform infrared spectroscopy (FTIR) measurements were performed by Thermo scientific FTIR 6700. The absorbance spectra were collected in the range of 400-4400 cm^{-1}. Each measurement was an average of 64 scans with a resolution of 4 cm^{-1}.

The chemical composition of organo-silica matrix of the sample was investigated with FTIR spectroscopy. Particular bands are recognized from the Si–CH$_3$ group: 780 cm^{-1}, 897 cm^{-1}, 1273 cm^{-1} [9]. The bands at 431 cm^{-1} and corresponding at 2978 cm^{-1} are assigned to C–H bond [167]. The band appearing at 699 cm^{-1} is recognized as coming from Si–C bond. Characteristic bands are assigned to Si–O–Si structure: 487 cm^{-1}, 1039 cm^{-1}, 1134 cm^{-1}. The band at 1595 cm^{-1} is recognized as characteristic for double bond C=C from the phenyl ring. The flat, broad band around 3400 cm^{-1} is attributed to a group of signals coming from stretching vibration of phenyl ring's C–H bonds, to the Si–OH groups and H$_2$O molecules presence [168].

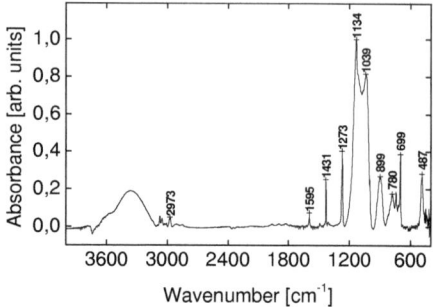

Fig. III.6.3. FTIR spectra of organo-silica matrix.

III.6.3.4. Thickness measurements

To perform thickness measurements of the film a part of the investigated material was removed by a scalpel in order to reveal SiO_2 substrate. AFM scan was started in the place where AFM tip scanned both the substrate and the material which thickness was to be measured.

The fast scanning direction was perpendicular to the line dividing the two materials giving rise to a clear step height difference in every measurement line. As a basis for the reliable height measurements the surface height histogram analysis was chosen (Fig. III.6.4) [169]. Clear two peaks in the surface height histograms can be noticed. Both peaks were fitted with Gauss functions and the peaks centre position were acquired from the fits. The step height equals the difference between the peaks centre positions. For the measured sample height of the deposited material was estimated to 144 nm. Height investigations require precise AFM levelling as a small misalignment between measuring axis and sample axis increases error in height measurements [170].

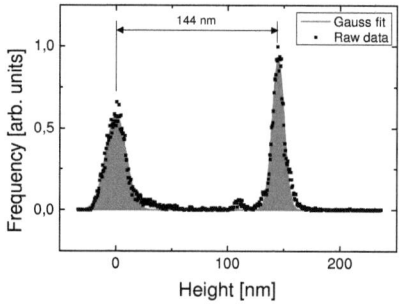

Fig. III.6.4. Surface height histogram of the region with both the substrate and the investigated material. Raw histogram data (black squares) were fitted by a Gauss functions (dark grey). The marked difference between peaks indicates the height of the deposited material.

III.6.3.5. CV measurements

CV measurements were performed on a MIS stack as described in experimental section. Fig. III.6.5 depicts results recorded at a frequency of 100 kHz at three different contacts positions of this sample. The CV curves show two aspects about the need of a strong quality improvement of the layer. The firs aspect is that the dispersion of the area capacitance indicates either a thickness and/or a porosity distribution inhomogeneity. The open porosity of this layer is estimated using ellipsometric porosimetry approximately 18 %. The pore size distribution is classified as mesoporous (mean pore size in the range of 2 nm-50 nm) [10]. Large hysteresis in the CV measurements indicates the presence of movable charges inside the layer. That phenomenon can be due to the Si–OH groups and H_2O molecules [118] in film pores (FTIR bands in the range of 3600-3200 cm^{-1}). This leads to a certain capacitance instability when the material is used in interconnects. In this stage of investigations we have not yet performed annealing steps which might lead to the avoidance of moveable charges inside the layers and, therefore, to an improved performance of the stack.

Besides the CV analysis the k-values were determined from the accumulation capacitance measurements. At two of the contacts we calculate a relative dielectric constant in the range of 2.3 to 2.5, on the other one it is about 20 % lower. Here the SiO_2 thickness of 100 nm and the thickness of 144 nm of the low-k layer as measured by AFM step measurements (see above) are taken into account. This result demonstrates that our fullerene based material may exhibit desired values for ULK application, however the layer quality should be improved by modified preparation methods.

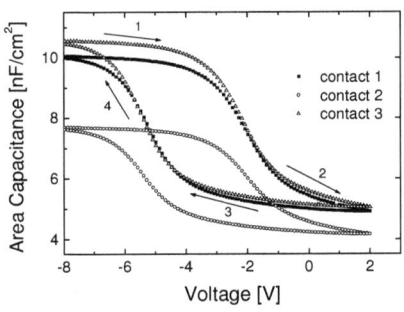

Fig. III.6.5. CV measurement on a MIS stack consisting of C_{60} derived compound. The measurements were taken at a frequency of 100 kHz at different contacts of the sample. The measurement direction is illustrated by the arrows.

III.6.4. Conclusions

Fullerene based low-k material has been obtained by means of the sol-gel method. Spectroscopic investigations confirmed, that the produced SiO_2 network is functionalized by the methyl and phenyl groups. For that reason the contributions to the carbon species existing within the film comes not only from the fullerenes but mainly from the phenyl rings and methyl groups. XPS analysis showed, that the overall amount of the carbon-like species within the sample is relatively high and the ratio of carbon, oxygen and silicon atoms within obtained film equals 2.7:1.9:1 respectively. One may presume that the low dielectric constant is due to the high concentration of the organic groups (methyl and phenyl) and porosity of the material. In order to reduce further the dielectric constant the concentration of the fullerene molecules should be significantly increased. Nevertheless the homogeneity of the layer needs further improvement. An outlook is to produce fullerene-based material while omitting the sol-gel procedure. In order to obtain better distribution of the fullerene molecules within the sample and its higher concentration as well the nitrogen containing siloxane precursor will be applied.

III.7. Investigations of the chemical and electrical properties of fullerene and 3-aminopropyltrimethoxysilane based low-k materials[5]

III.7.1. Introduction

The improvement of the microelectronic device performance has been achieved mainly by the continuous miniaturization of the feature size what allows higher packing density [10]. However, while scaling the microelectronic devices dimensions the overall resistance-capacitance (RC) delay increases and becomes the limiting factor of the electrical and functional performance. The challenge connected with the RC delay reduction has been mitigated by the replacement of the Al/SiO_2 on-chip interconnect stack by Cu and low dielectric constant (low-k) materials. Introduction of the low-k materials and Cu interconnects allows more over to reduce the power consumption and cross-talk of the microelectronic devices [10,11,171].

---

[5] J. Klocek, K. Henkel, K. Kolanek, K. Broczkowska, D. Schmeißer, M. Miller, E. Zschech, Studies of the chemical and electrical properties of fullerene and 3-aminopropyltrimethoxysilane based low-k materials, Thin Solid Films 520 (2012) 2498-2504

Since fullerenes exhibit a number of unique properties that are valuable in many fields of science, they attract a great scientists' attention and big efforts have been made in order to obtain fullerene containing nanocomposite materials [54,172]. The presence of the fullerenes within the low-k material can improve its mechanical properties [60], and according to theoretical calculations it can also significantly decrease the value of its dielectric constant [61].

In this chapter the first step leading to the preparation of APTMS based ultra-low-k material including fullerene cages is described. The main purpose of the first investigations was to characterize the chemical stability of the material after exposure to the air and estimate the influence of fullerenes replacement by its better soluble derivatives [6,6]-phenyl-C_{61}-butyric acid methyl ester (PCBM) on the dielectric constant of the films. Hence, I have prepared fullerene based material consisting of 3-aminopropyltrimethoxysilane (APTMS) and C_{60} molecules by spin coating. Afterwards another similar sample has been produced while using PCBM instead of C_{60} in order to minimize the amount of the clusters within the low-k film.

APTMS has been used as a basic material for the samples preparation. The fullerene functionalization with this agent has already been reported in the literature [14,15]. The reason for using APTMS in this experiment is not only its high affinity to the fullerene molecules but also fact that this agent functionalizes effectively the silicon surface since APTMS exhibits properties of forming self-assembled monolayer [119,134,173–176]. APTMS gives an opportunity to introduce aminopropyl groups into the framework of the final material. Functionalization of the mesoporous silica structure with organic groups originating from organosiloxanes opens a new horizons for many fields of science [177].

Silicon has been chosen as a substrate for this experiment because of its wide applications in the field of microelectronics devices. For that reason, there is an obvious need to understand the interactions between silicon single-crystalline substrates and the deposited thin film material [12]. The knowledge of the electronic structure of the semiconductor-organic interface is a fundamental factor that influences device performance improvement [178].

The chapter is organized as follow: Based on the XPS the stability of the hybrid material films composed of the fullerene species incorporated into the APTMS based siloxane matrix spin-coated on the silicon substrate is discussed. Subsequently, the studies of the similar material produced after the C_{60} replacement by PCBM are described. In the next section, we present and discuss an example of the prepared films' thickness measurement by means of atomic force microscopy (AFM). The final part describes samples' electrical characterization by capacitance-voltage (CV) method and obtained results. Finally, a summary of the relevant conclusions is given.

III.7.2. Experimental

III.7.2.1. Materials and substrates

In this work C_{60} (99,5 %), PCBM (99,5 %) and APTMS (97 %) obtained from Sigma-Aldrich have been used. A substrate for the samples preparation was Si (100).

III.7.2.2. Samples preparation

Fig. III.7.1 illustrates schemes of all described in this chapter samples. The C_{60} and PCBM concentrations within the films were calculated in weight percent.

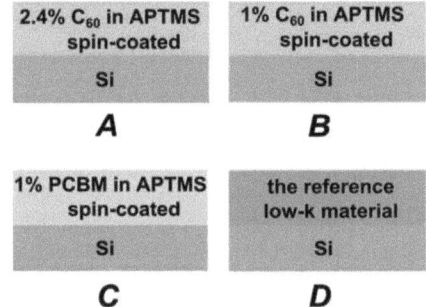

Fig. III.7.1. The schemes of the investigated samples: *A)* 2.4 % C_{60} solution in APTMS spin-coated on the Si surface (referred in the text as sample *A*), *B)* 1 % C_{60} solution in APTMS spin-coated on the Si surface (sample *B*), *C)* 1 % PCBM solution in APTMS spin-coated on the Si surface (sample *C*), *D)* low-*k* film obtained by the sol-gel method used as a reference sample (sample *D*, see also chapter III. 6).

After 10 minutes of the ultrasonification in isopropanol and then in distilled water, the Si substrates were dried in nitrogen stream. The hydroxylation of the wafers was achieved by the immersion in piranha solution (a mixture of 7 : 3 (v/v) 98 % H_2SO_4 and 30 % H_2O_2) at 120°C for 1.5 h. The substrates were then rinsed with distilled water and isopropanol several times and dried in nitrogen stream.

The materials including C_{60} and PCBM were produced by stirring with APTMS in glove box under argon atmosphere for 24 h. Films were deposited onto the substrates by spin-coating technique at a speed of 6000 rpm for 15 seconds inside glove-box. The C_{60} containing sample designed for the stability measurements were prepared in the similar way but in ambient conditions.

The reference film (sample *D*) was obtained by sol-gel method (see chapter III. 6). A basic sol solution was prepared by mixing phenyl triethoxysilane : methyl trimethoxysilane : ethanol : hydrochloric acid in molar ratio 1.00 : 1.22 : 7.77 : 1.03 × 10^{-3} respectively. The detailed preparation procedure of this sample is reported in [179].

Additionally, samples *B* and *C* were used for CV measurements. Therefore, MIS structures were prepared by evaporation of silver contacts on top of the samples using a shadow mask. The diameters of the contact areas ranged between 400 µm and 700 µm. Two stacks corresponding to sample *B* were prepared for electrical measurements.

III.7.3. Results and discussion

III.7.3.1. XPS measurements

XPS is a common tool applicable to the determination of the materials chemical composition. The peak positions in the XPS spectra in respect to the binding energy provide information concerning the chemical state of the particular atoms within the measured films [180]. The analysis of the peak intensities is used in order to perform quantitative determination of the components in the material [27].

III.7.3.1.1 Investigations on the stability of sample *A*

In this part the influence of ambient conditions onto the material composed of APTMS and C_{60} is reported. Sample *A* was investigated twice: after 24 h of drying in the air (further referred as sample *A*, Fig. III.7.2(a-d)) and then after 10 days of the storage in the ambient conditions (sample *A'*, Fig. III.7.2(e-h)). The results of the detailed spectra analysis are summarized in Table III.7.1.

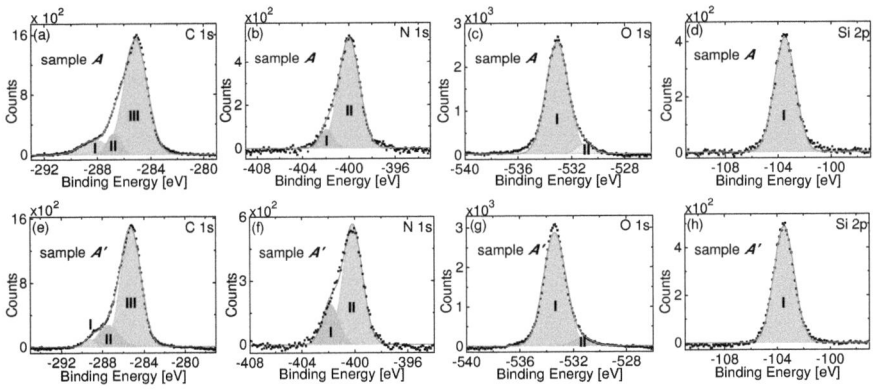

Fig. III.7.2. XPS spectra of the sample *A*. Top row (a-d) represents the sample *A* measured 24 h after the preparation (sample *A*), bottom row (e-g) depicts the same sample measured 10 days after the preparation (sample *A'*).

Table III.7.1. Peak deconvolution of the XPS spectra of sample *A*: obtained after 24 h (sample *A*) and ten days after the preparation (sample *A'*) respectively.

Region	Peak Number		Position [eV]		Percentage of the main peak area [%]		Assignment	
	Sample A	Sample A'	Sample A	Sample A'	Sample A	Sample A'	Sample A	Sample A'
C 1s	I	I	288.2	289.2	12	2.8	oxidized C species [63,121]	NHCOO⁻ [119]
	II	II	286.8	287.4	8.7	15.9	C–N sp^3[121], C–N [63]	C–O[63], C–N[181]
	III	III	285.1	285.3	79.3	81.3	carbon sp^3[24]	carbon sp^3[24]
N 1s	I	I	401.9	401.9	9.4	21.8	-NH$_3^+$ [12,119]	-NH$_3^+$ [12,119]
	II	II	400	400.2	90.6	78.2	-NH$_2$ [6,121]	-NH$_2$ [6,121]
O 1s	I	II	533	533.4	94.1	95.8	-C-O [127], -Si-O [128]	-C-O[71]
	II	---	531	531.5	5.9	4.2	-N=C=O [127]	-N=C=O [127]
Si 2p	I	I	103.5	103.5	100	100	O-Si-O/Si-OH[128]	O-Si-O/Si-OH[128]

The C 1s overall signal deconvolution shows that the main peak corresponding to the sp^3 hybridized carbon species (signal III in Fig. III.7.2a and 2e) appears at around 285.1 eV and 285.3 eV for sample *A* and *A'*, respectively. For the pure APTMS the main feature exists at around 286 eV and corresponds to the propyl chain of the molecule [119]. Therefore one may presume that the carbon species of the fullerenes shifted the main carbon feature of the XPS spectrum. In the C

1s spectrum of sample A (Fig. III.7.2a) one observe two smaller features at around 286.8 (peak II) and 288.2 eV (peak I) that correspond to the C–N and C–O species, respectively. The C 1s middle feature (signal II) of the sample A' (Fig. III.7.2e) is quite broad because it consists of two carbon states: C–N and C–O. In the C 1s spectrum of the sample A' we observe also an additional characteristic peak at around 289.2 eV (signal I) which is assigned to the carbamate group (NHCOO⁻) that is formed due to the reaction between –NH_2 group and CO_2 from the ambient. This process is confirmed by the significant increase of the corresponding feature of the –NH^{3+} group in the N 1s spectrum recorded for the sample A' (Fig. III.7.2f, signal I) in comparison to sample A (Fig. III.7.2b, signal I). This observation suggests that the APTMS and C_{60} based films could be affected by CO_2 presence in ambient. Details regarding the interactions of the APTMS based surface with the CO_2 from the air are described by reactions III.5.1 and III.5.2 [119].

Since XPS spectroscopy enables a quantitative determination of the elements contained in the measured sample [27] after the spectra evaluation the quantitative analysis of the investigated films was performed. The spectra normalization was achieved by dividing the observed relative peak areas by atomic and instrument sensitivity factors equal to 0.711, 0.296, 0.477, and 0.339 for O, C, N, and Si, respectively [25]. As a result of the analysis the following composition in atomic percent has been obtained: C=52.3 %, N=9.7 %, O=29.1 %, Si=8.9 % ($C_{5.9}N_{1.1}O_{3.3}Si$) for the sample A and C=51 %, N=9.5 %, O=29.8 %, Si=9.7 % ($C_{5.3}NO_{3.1}Si$) for the sample A'. The stoichiometry of an unhydrolyzed APTMS monomer (excluding H atoms) corresponds to C_6NO_3Si, in comparison, the "ideal" poly[(aminopropyl)siloxane] polymer has a stoichiometry of $C_3NO_{1.5}Si$ [119]. The decrease of the overall C amount within the film A and A' in comparison to the unhydrolyzed monomer suggests that the APTMS monomers have partially undergone the condensation reactions during the spin-coating and drying and some –CH_3 species were removed from the material. After ten days of the sample storage we observed further progress of the condensation process. This could be a hint that in order to obtain films with more stable chemical composition the samples ought to be annealed straight after the preparation. However, taking into account that the atomic ratio of the Si and N within the described film remains close to 1 we can draw conclusions that the aminopropyl species were not removed from the siloxane matrix.

III.7.3.1.2. Investigations of the samples B and C

The main aim of this part of the experiment was to estimate how the replacement of C_{60} by PCBM influences the chemical composition of the final material. For this purpose samples B [Fig. III.7.3(a-d)] and C [Fig. III.7.3(e-h)] have been investigated after 24 hours of the exposure to air. The results of the detailed spectra analysis are summarized in Table III.7.2.

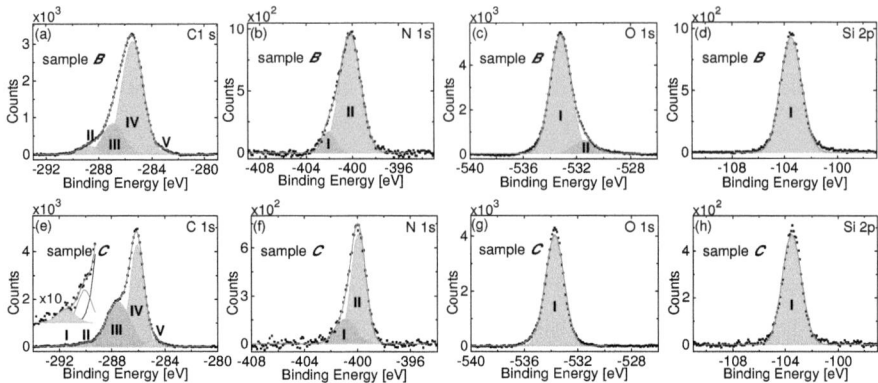

Fig. III.7.3. XPS spectra of the samples *B* and *C*. Top row (a-d) represents the sample *B*, bottom row (e-g) depicts the sample *C*. The inset in (e) shows the magnified region I.

First of all, while analysing the C 1s XPS spectrum of sample *C* (Fig. III.7.3e) one may notice the characteristic fullerene feature attributed to the π-π* electron transition (signal I) [63,183]. In contrast, the C 1s spectrum of the sample *B* (Fig. III.7.3a) reveals no characteristic C_{60} shake-up features, what is the proof of very low C_{60} concentration within that film. The C 1s spectrum of the sample *C* contains also, contrary to that of the sample *B*, one small feature around 284.8 eV (signal V in Fig. III.7.3a) which corresponds to the C–C species within the C_{60} cage [63]. The main feature of the sample *B* (signal IV in Fig. III.7.3e) appears around 285.4 eV. As described in section 3.1.1, the main APTMS feature coming from the propyl group is normally placed at around 286 eV [121], what is fulfilled for the sample *C* where the main C 1s peak is at around 286.1 eV (see Fig. III.7.3e, peak IV). This allows supposing that the carbon main signal of the sample *B* placed around 285.4 eV is shifted to the lower binding energy direction due to the C_{60} contribution which was too low to distinguish a particular fullerene feature in deconvolution process.

Table III.7.2. Peak deconvolution of the XPS spectra of samples B and C.

Region	Peak Number		Position [eV]		Percentage of the main peak area [%]		Assignment	
	Sample B	Sample C	Sample B	Sample C	Sample B	Sample C	Sample B	Sample C
C 1s	----------	I	----------	291.5	----------	1	----------	C1s shake up satellite [12]
	II	II	288.8	290.1	5.2	1.8	Oxidized C-species C=O[63]	-NHCOO⁻ [119]
	III	III	286.9	287.5	20.1	400.5	C-N [63]	C-N sp³[121]
	IV	IV	285.4	286.1	73.3	52.9	Carbon sp³[70]	-(CH$_2$)$_3$- [119]
	V	V	----------	284.8	----------	3.8	----------	C-C species from PCBM
	VI	----------	283.1	----------	1.4	----------	Si-C [73][182]	----------
N 1s	I	I	402	400.9	9.6	23.4	-NH$_3^+$ [119,12]	-NH$_3^+$ [119,12]
	II	II	400.2	399.9	90.4	76.6	-NH$_2$[6,121]	-NH$_2$[6,121]
O 1s	I	II	533.2	533.7	89.5	97.7	-C-O[127], -Si-O[128]	-C-O[71]
	II	----------	531.4	----------	10.5	----------	-N-C=O[127]	----------
Si 2p	I	I	103.5	103.5	100	100	O-Si-O/Si-OH[128]	O-Si-O/Si-OH[128]

The quantitative analysis of the film B reveals following surface composition in atomic percent: C=51.5 %, N=9.3 %, O=29.2 %, Si=10 % (C$_{5.9}$N$_{1.1}$O$_{3.3}$Si). The atomic ratio obtained for the sample B is in good agreement with those obtained for the sample A. Although the concentration of the fullerenes used for the film preparation equalled 2.4 % and 1 % for the sample A and B, respectively, the quantitative XPS analysis revealed that after 24 h of the exposure to air the atomic ratios for both films were almost identical. This proves that the C$_{60}$-APTMS solutions used for the samples A and B preparation were already saturated and it is not possible to increase the fullerene amount within these types of films. Following our earlier expertise (described in chapter III.6 sample D analysis), while increasing the fullerene concentration the dielectric constant of the material ought decrease due to the higher overall amount of pores and carbon species within the film. Sample D used in this experiment as a reference contained high amount of carbon species and was porous what allowed to achieve its very low dielectric constant (from 2.3 to 2.5), although the C_{60} amount within the film was quite low (0.3 %). For that reason I decided to further increase the amount of the fullerene species within the film by replacing C_{60} by its better soluble derivative: PCBM (sample *C*). XPS quantitative analysis revealed that after 24 h of exposure to the air film's *C* composition in atomic percent was following: C=69.4 %, N=5.2 %, O=20.4 % and Si=5 % ($C_{13.9}NO_4Si$). Taking into account that in the preparation of samples *B* and *C* the identical C_{60} and PCBM concentration (1 %) was used one may conclude, that the purpose of increasing the amount of the C_{60} species within the film succeeded. For the sample *C* the overall amount of the carbon species equalled 69.4 %. Sample *C* contained also higher amount of the O species in comparison to the sample *B* since PCBM molecule is also a source of this element.

III.7.3.2. Thickness measurements

We used AFM to determine thickness (*d*) of the deposited materials in order to calculate dielectric constant from CV measurements (Eq. III.7.3). As all investigated materials were soft, to perform thickness measurements of the deposited films a part of the investigated material was carefully removed by a scalpel in order to expose the Si/SiO_2 substrate. We started the scan in the place where AFM tip sensed both the substrate and the material which thickness was to be measured. A typical example of the thickness measurement for the sample *C* is presented in Fig. III.7.4(a). By setting the fast scanning direction perpendicular to the vertical line dividing the two regions a clear step height difference in every measurement line was recorded [Fig. III.7.4(b)]. AFM thickness investigations require precise surface leveling as a small misalignment between measuring axis and sample axis increases the error in height measurements [184]. After correct surface leveling the two regions are fitted by a least square line in which the slope is constrained to zero. From the fit we rejected the edges of the image and the step as both are prone to AFM artifacts [170,185]. After this procedure the difference between individual intercepts of the lines equal to 65 nm is the thickness of the investigated material. The usage of this procedure for step height measurements is however time consuming, as every line should be analyzed independently. A good alternative for the reliable height measurements is the surface height

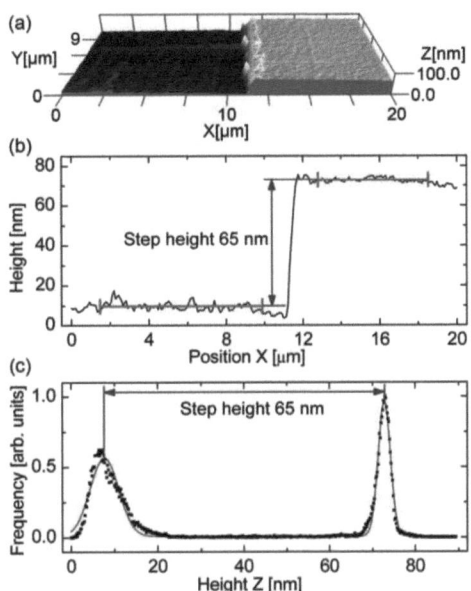

Fig. III.7.4. NC-AFM thickness measurements of the sample C at randomly chosen position. (a) Surface topography of the two regions: Si/SiO$_2$ substrate (dark color), and 1 % PCBM in APTMS spin-coated surface (light color). (b) Typical profile of the step with linear fits to the two regions. (c) Surface height histogram from the whole image with two Gaussian fits.

histogram analysis [Fig. III.7.4(c)] [186]. A height histogram has the advantage of a simultaneous analysis of all the pixels of the image what in turn is manifested as an averaging effect in which the influence of AFM distortions are not pronounced. For a stepped surface two well pronounced peaks in the surface height histogram are observed. Each peak was fitted using Levenberg–Marquardt least-square algorithm, with a Gauss function and the peaks center positions were acquired from the fits. The step height equals the difference between the peaks centre positions [187,188]. From the analysis of the Fig. III.7.4(c) we found the height of the deposited material to be equal to 65 nm. In general, the thickness of our films varied between 53 nm and 74 nm. For all samples we measured the thickness at least at 3 different positions distributed over the surface near the evaporated contacts prepared for CV investigations. This allowed us to apply statistics and within each sample we estimate a height percentage deviation to be of the order of ± 10 %, which is in good agreement with capacitance fluctuation of 9 %. This is an indication that spreading of capacitance in accumulation is mainly caused by material thickness deviation. The thickness measured by AFM is the major source of error for the determination of the dielectric constant which error we assume to be in the range of ± 12 %

As the width of the Gaussian peaks is equal to the surface height standard deviations, we may individually determine the RMS roughness of both materials [157,187]. We obtained surface height standard deviations in the order of 3.3 nm and 1.4 nm for the substrate and 1 % PCBM in APTMS material, respectively.

III.7.3.3. CV measurements

The CV measurements of samples B and C at three different contact areas at each sample are shown in Fig. III.7.5 (a) and (b). The area capacitance in accumulation shows a relative constant value for the three contacts on both samples. Compared to former investigations on sol-gel films with C_{60} components (chapter III.6) where a spreading of the area capacitance of approximately 25 % was found we observe in these films prepared by spin coating a spreading of only 9 %. This indicates that the film homogeneity of the spin coated films seems to be better than that of the sol-gel film.

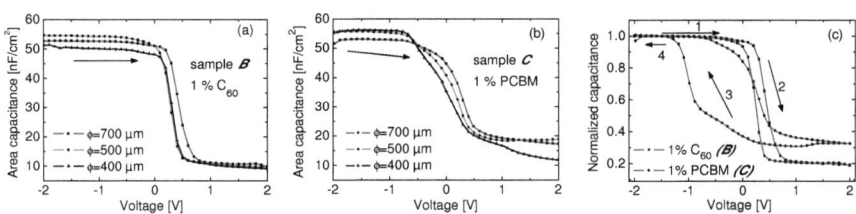

Fig. III.7.5. CV measurements recorded at 1 MHz on the samples B and C. Figure part (a) depicts the measurements on sample B at three different contacts with various diameters as given in the legend, while the same is shown in figure part (b) for sample C. In part (c) the hysteresis during the measurement on both samples is pictured. The arrows are used in order to illustrate the measurement direction.

However, we still observe a huge hysteresis inside the CV loop as shown in Fig.III.7.5 (c). This is much more pronounced in the PCBM film (sample C). At this stage of investigations we have not yet performed annealing steps which might lead to the avoidance of moveable charges inside the layers corresponding to the hysteresis in the CV loops. Furthermore, a minuscule amount of impurities like absorbed water in the film may ruin the effort of low-k integration [118] caused by the high polarity of water and consequently its high dielectric constant in the range of 80 (at room temperature and 0.1 MPa) [189] pointing out the necessity of post deposition annealing routine.

Nevertheless, the improved homogeneity of the spin coated films allows a plot of the normalized capacitance taken in accumulation (C_{acc}) versus the contact area (A) as depicted in Fig III.7.6. In order to determine the k-value of the films the plate capacitor approximation was applied. Writing

$$C_{acc}\frac{d}{\varepsilon_0} = kA, \qquad (III.7.3)$$

where d is the film thickness, and ε_0 is vacuum permittivity, it can be easily seen that the dielectric constant k of the film can be directly determined by the slope of a linear fit to the data. Here the thickness determined near each related contact used for the CV measurements on the samples was applied in Equ. (III.7.3). The lower slope of the fit to the data of sample C containing 1 % of PCBM shows that this sample exhibits a smaller k-value equal to 3.4, in contrast to sample B where we obtained $k = 4.1$. Here we have to mention that on the Si substrate a residual SiO_2 layer was still existent after the cleaning procedure as revealed in the XPS investigations (not shown). Assuming a SiO_2 thickness of 1 nm to 2 nm the resulting influence on the permittivity value of our layers is in the range of 1.6 % to 3.2 %. Therefore, the SiO_2 layer was neglected in the calculations.

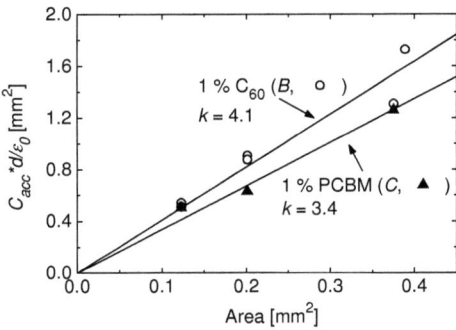

Fig. III.7.6. Normalized accumulation capacitance versus contact area of samples B and C and linear fits performed according to Equ. (III.7.3).

Concerning these results we can conclude that the usage of a film containing the better soluble fullerene derivative PCBM is leading to lower k-values compared to films produced by solutions of pure C_{60}.

III.7.4. Conclusions

APTMS based low-k material including fullerene cages were synthesized and investigated. With XPS spectroscopy, an analysis of the time-related chemical processes within the prepared films exposed to the ambient conditions was performed. It was confirmed that the fullerene containing APTMS monomer spin-coated on the substrate partially undergoes the condensation reaction loosing $-OCH_3$ and $-CH_3$ species and forming Si–O–Si network. However, quantitative XPS analysis of the described samples reveals that the ratio of the N and Si atoms within the films remains close to 1. This result proves that the aminopropyl groups are not removed from the siloxane matrix. The 1 % C_{60} APTMS solution is already saturated, and films prepared by spin-coating contain about 51 % of the carbon species. This value may be increased up to about 69 % by the replacement of the C_{60} with PCBM molecules. Electrical measurements showed that the increase of the overall carbon species amount within the film reduces the dielectric constant of the prepared material. Large hysteresis of the CV curve obtained for the PCBM containing sample suggests the presence of the movable charges within the film that may be by-products of the chemical reactions between the components. Adsorption of the small amount of water and other impurities that could influence the hysteresis formation cannot be excluded. Summarizing, a significant homogeneity improvement of the films was obtained due to the spin-coating preparation, in comparison to films prepared previously by the sol-gel method (described in details in chapter III.6). At the same time, the fullerene species concentration was significantly increased. The prepared thin film material has a low dielectric constant. Further improvement (water elimination, replacing fullerenes by its derivatives, additional dopants application) will be needed to obtain high quality ultra-low-k films.

III.8. Influence of the fullerene derivatives and cage polyhedral oligomeric silsesqiuoxanes on 3-aminopropyltrimethoxysilane based hybrid nanocomposites chemical, morphological and electrical properties[6]

III.8.1. Introduction

Hybrid nanocomposites are new functional compositions which, depending on the manufacture route, may possess truly extraordinary properties. In general, they are a dispersion of nanoparticulates into a macroscopic matrix while their structures include both organic and inorganic groups. The nanoscale reinforcement results in new phenomena which contribute to material properties [190]. Different types of the nanocomponents improve plenty of the material attributes like optical, dielectric, electrical, mechanical etc. Since they can be considered as a promising and attractive technological platform for electronic, optical, magnetic, and biomedical applications, hybrid nonocomposites are of great interest of the nanotechnology [57].

In the present chapter I used 3-aminopropyltrimethoxysilane (APTMS) as a macroscopic matrix since, as it was proved in the previous chapter, it may be successfully used for thin low-k film preparation in a spin-coating process.

Since fullerene molecules are very resistant both from physical and chemical point of view, their incorporation into the matrix of the chosen material can significantly improve the properties of the final material. However, there are some reasons such as ability to form clusters and low fullerenes solubility that forces the replacement of C_{60} by its better soluble derivatives [107,191]. Thus, thin APTMS films were modified with the [6,6]-phenyl-C_{61}-butyric acid methyl ester (PCBM) molecules, regarding the experience described in chapter III.7.

In order to improve the properties of the final films for possible low-k application in this work the nanocomposite materials were produced using not only PCBM nanodopants but also incompletely condensed frameworks possessing a hybrid inorganic-organic, three-dimensional structure cage polyhedral oligomeric silsesqiuoxanes (POSS). The very interesting feature of POSS is their hybrid architecture composed of the inner inorganic framework $(SiO_{1.5})_x$, covered externally by organic substituents what makes them compatible with polymers [192]. Organic groups, that can be specially designed to be either nonreactive or reactive [192], allow to control the dielectric

---

[6] J. Klocek, K. Kolanek, K. Henkel, E. Zschech, D. Schmeisser, Influence of the fullerene derivatives and cage polyhedral oligomeric silsesqiuoxanes on 3-aminopropyltrimethoxysilane based hybrid nanocomposites chemical, morphological and electrical properties, (submitted).

constant of the material and its many other properties ranging from the thermal stability, hydrophobicity, refractive index, oxidation resistance, surface hardening to the optical clarity [57,193]. Moreover, organic substituents may provide the opportunity to serve as spacer between Si–$O_{1.5}$ linkages introducing a well-modulated porosity within the material [57]. This is a strongly desired feature in the case of ultra-low dielectric constant materials [61].

Since POSS incorporation into the polymeric material can lead to many substantial varieties and improvements in final compositions, investigations regarding this topic has attracted much research interest in the past few years [193–195]. The common approaches for hybrid POSS consisting material manufacturing are blending or covalently bonding POSS to a polymer matrix, which usually results in phase separated composites [196]. The present chapter describes the first step leading to low-*k* material reinforced with POSS and PCBM. Thus, the incorporation process was performed without using any external initiator in order to estimate the compatibility of applied components and the influence of the POSS molecules on the permittivity of the final composition. This chapter focuses on the chemical interactions between the material components, on the particles and matrix morphology (particle size and arrangement), and on the permittivity of the final compositions investigated by X-ray photoelectron spectroscopy (XPS), atomic force microscopy (AFM), and capacitance-voltage (CV) methods, respectively.

III.8.2. Experimental

III.8.2.1. Materials and substrates

In this work used PCBM (99.5 %), APTMS (97 %) and Tris(dimethylvinylsilyloxy)-POSS (further referred as POSS in the text), obtained from Sigma-Aldrich have been used. As a substrate for the samples preparation Si (100) has been chosen because of its fundamental application in the field of microelectronic devices [12] and in order to build an appropriate for the CV measurements MIS structure. The phosphorous doped n-type Si wafers with a resistivity of 1-5 Ωcm were purchased from Crystec (Berlin, Germany).

III.8.2.2. Sample preparation

Fig. III.8.1 presents the schemes of the investigated samples. The concentrations were calculated in weight percent.

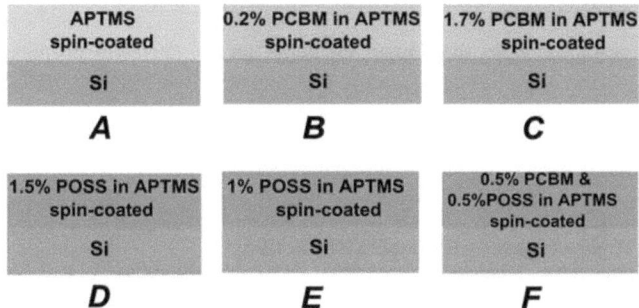

Fig. III.8.1. The schemes of the investigated samples: *A* APTMS spin-coated on the Si surface (referred in the text as sample *A*), *B* 0.2 % PCBM solution in APTMS spin-coated on the Si surface (sample *B*), *C* 1.7 % PCBM solution in APTMS spin-coated on the Si surface (sample *C*), *D* 1.5 % POSS solution in APTMS spin-coated on the Si surface (sample *D*), *E* 1 % POSS solution in APTMS spin-coated on the Si surface (sample *E*), *F* APTMS solution containing 0.5 % PCBM and 0.5 % POSS spin-coated on the Si surface (sample *F*).

After the ultrasonification in isopropanol for 10 minutes and then in distilled water, the Si substrates were dried in nitrogen stream. The surface of the wafers were hydroxylated by the immersion in piranha solution (a mixture of 7 : 3 (v/v) 98 % H_2SO_4 and 30 % H_2O_2) at 120 °C for 1.5 h. The substrates were then rinsed with distilled water and isopropanol several times and were dried afterwards in nitrogen stream.

For the sample *A* preparation pure APTMS has been used. The materials of samples *B*, *C* and *E* were produced from a dispersion of PCBM and POSS particles, respectively, in the APTMS monomer. Sample *D* was obtained in a similar way like sample *E*, but the solution has been further diluted in APTMS after one week of its storage under argon atmosphere. The material of sample *F* was obtained by mixing solutions of 1 % POSS in APTMS and 1 % PCBM in APTMS in a proper ratio. The samples were produced in the form of thin films by spin-coating at a speed of 6000 rpm for 15 seconds in glove box conditions.

Additionally, the samples *A*, *B*, *E* and *F* were used for CV investigations. In order to prepare MIS structures, silver contacts were evaporated on top of the layer stacks shown in Fig. III.8.1 using a shadow mask with diameters in the range of 400 μm to 800 μm. In the CV results a sample of 1 % PCBM in APTMS (sample *C* described in chapter III.7) is added for comparison, indexed in the following as sample *G*.

III.8.3. Results and discussion
III.8.3.1. XPS investigations on the dopant influence on the chemical properties of APTMS films

The main motivation of this experimental part is the investigation on the strength of the chemical interactions between APTMS, PCBM and POSS, depending on the doping concentration.

III.8.3.1.1. PCBM nanodopants within the APTMS films

Here the influence of the PCBM nanodopants on the chemical properties of the APTMS films will be discussed. The preparation of the samples composed of APTMS and PCBM of different concentration proved that those nanodopants can be the crucial factor for obtaining high quality spin coated films. This was already evident after the macroscopic observation of the samples *A*, *B* and *C*. In the case of the sample *C* the concentration of the PCBM was too high (1.7 %) to obtain a homogenous film of uniform thickness. Furthermore, in this case the PCBM agglomerated into structures which were too large to investigate the surface by AFM. On the other hand the APTMS solution containing 0.2 % of the PCBM turned out to be excellent for the spin-coating process. Although the conditions of the coating process were the same for each sample, only in the case of the sample *B* the film could not be seen visually and the evidence of its existing could be proven only by XPS and AFM investigations.

Here are report the investigations of the samples *A* (Fig. III.8.2a-d), *B* (Fig. III.8.2e-h), and *C* (Fig. III.8.2i-l), which were measured 24 h after the deposition and storage in ambient conditions. The detailed peak analysis is given in Table III.8.1. Corrections for the energy shift were accomplished assuming a binding energy of 103.5 eV for the Si 2p peak of the Si^{+4} species [119].

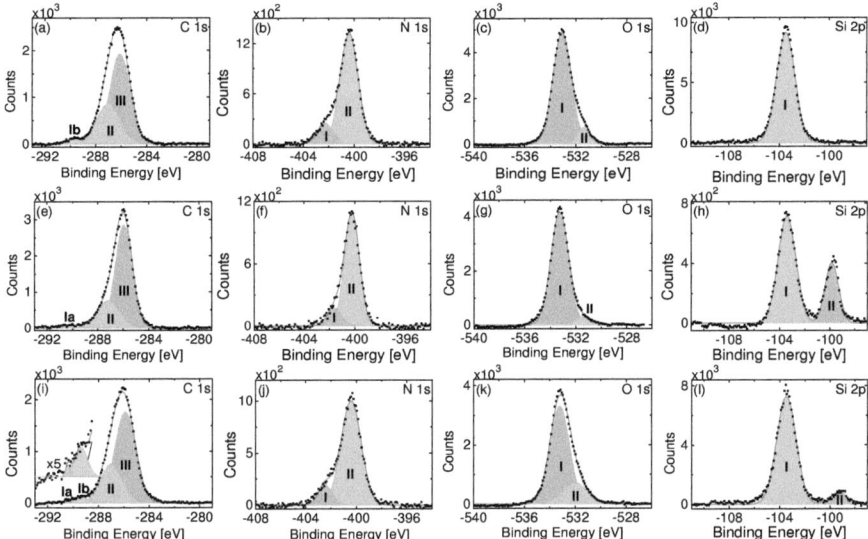

Fig. III.8.2. XPS spectra of the samples *A*, *B* and *C*. Top row (a-d) represents the sample *A*, middle row (e-h) depicts sample *B*, bottom row (i-l) sample *C*, respectively. The following core level spectra are drawn in the same order for every sample: C 1s, N 1s, O 1s, Si 2p. The inset in plot (i) shows the magnified signals Ia and Ib.

First the C 1s core level spectra of the samples *A*, *B*, and *C* is discussed. Since the XPS spectra consist of a mixture of a few chemical states of the investigated element, extracting every particular chemical state in the decomposition process is a quite challenging task, especially when the measured XPS spectra exhibit a low resolution. For example, the signals corresponding to the NHCOO⁻ species of APTMS and the shake-up satellites from PCBM overlap each other in the C 1s region of the XPS spectra (Peaks Ia and Ib in Fig. III.8.2a, 2h, and 2i). A similar situation occurs in the case of the peaks coming from the C–O and C–N species (Peaks II in Fig. III.8.2a, 2e, and 2i). However, while analyzing shape and central position of such hybrid peaks one may draw valuable conclusions concerning the contribution of the particular species within the investigated surfaces.

Table III.8.1. Peak decomposition of the XPS spectra of samples A, B and C.

Region	Peak number	Position [eV]			Assignment		
		A	B	C	A	B	C
C 1s	I a	-------	-290.2	-290.4	----------	NHCOO⁻ [119] &π-π* shake-up[120]	NHCOO⁻ [119] &π-π* shake-up[120]
	I b	-289.7	-------	-289.4	NHCOO⁻ [119]	----------	NHCOO⁻ [119] &π-π* shake-up[120]
	II	-287	-287.2	-287	C–N sp^3 [121]&C–O[63]	C–N sp^3 [121]&C–O[63]	C–N sp^3 [121]&C–O[63]
	III	-286.1	-286	-285.9	–(CH$_2$)$_3$– [119]	–(CH$_2$)$_3$– [119]	–(CH$_2$)$_3$– [119]
N 1s	I	-402.3	-401.6	-402.5	–NH^{3+}[119]	NHCOO⁻ [119]	–NH^{3+}[119]
	II	-400.4	-400.2	-400.4	–NH$_2$[119]	–NH$_2$[119]	–NH$_2$[119]
O 1s	I	-533.1	-533.3	-533.2	–C–O[127], –Si–O[128]	–C–O[127], –Si–O[128]	–C–O[127], –Si–O[128]
	II	-531.4	-531.4	-532	–N–C=O[127]& C=O[126]	–N–C=O[127] & C=O[126]	–N–C=O[127]& C=O[126]
Si 2p	I	-103.5	-103.5	-103.5	O–Si–O/Si–OH[128]	O–Si–O/Si–OH[128]	O–Si–O/Si–OH[128]
	II	-------	-99.5	-99.4	-------	Si^0[197]	Si^0[197]

While analyzing the C 1s core level spectra of sample C one can notice weak but at the same time characteristic PCBM satellite features in the range between -291 eV and -289 eV (signal Ia at -290.4 eV and Ib at -289.4 eV in Fig. III.8.2i). According to the literature, the positions of the C$_{60}$ shake-up features are normally observed at -291.2 eV and -289 eV [120]. The center positions of those peaks are a bit shifted in the C 1s data of sample C. As it was mentioned above, this is due to the presence of the carbamate species formed as a result of the amine group interactions with ambient CO$_2$ molecules [119]. Nevertheless, the presence of those characteristic fullerene cage shake-up features proves that a relative large number of π-bonds of the PCBM molecules remained intact. Therefore one may suppose that the particular PCBM molecules were not damaged by the interactions with the APTMS. For a very low PCBM concentration in the APTMS (sample B) it is very difficult to distinguish the particular shake-up features in the C 1s XPS spectrum what may be explained by the low PCBM concentration within the film (0.2 %) combined with the interactions of some of the π-bonds of the fullerene cage with the amine group of the APTMS. In comparison, the C 1s XPS spectrum of the pure APTMS (sample *A*, Fig. III.8.2a) exhibits only one contribution in this range, coming from the carbamate carbon signal at around -289.7 eV.

The main contributions to the features II within the C 1s data of the discussed samples come from the C–N and C–O species. As mentioned above these contributions are overlapping each other. Thus, those features centered at around -287 eV are quite broad (FWHM equals 2.18, 2.16 and 2.00 for the samples *A*, *B*, *C* respectively).

The highest features obtained in the C 1s XPS spectra decomposition (signals III) are corresponding to the carbon atoms from the propyl groups of the APTMS molecules [119], [198]. The peaks are centered at -286.1 eV, -286 eV and -285.9 eV for the samples *A*, *B*, *C* respectively. However, also those signals contain the contribution of different carbon states. For the samples *B* and *C* the slight shift (0.1 eV and 0.2 eV) of the main peaks to lower binding energies is attributed to the contribution of the C–C of the PCBM molecules.

Next the N 1s XPS spectra of the samples *A*, *B* and *C*, which were decomposed into two features (signals I and II in Fig. III.8.2b, 2f, and 2j) are discussed. The higher energy signals (peak I) of the samples *A* and *C* are attributed to the positively charged nitrogen atom [119], while the main contribution to the feature I of sample *B* is attributed to the carbamate nitrogen originating from the NHCOO⁻ group [119]. Both species formed as a result of APTMS interactions with the ambient CO_2 regarding the reactions III.5.1 and III.5.2 [119].

The low energy features of all samples presented in this part are attributed to the free amine groups of the APTMS molecules [119]. The number of the amine groups that underwent chemical interaction with the PCBM molecules was apparently very low. Therefore, in the N 1s spectra we cannot observe significant features corresponding to the secondary amine that would form as a result of the reaction with the fullerene cage.

In the O 1s core level spectra two features have been obtained (Fig. III.8.2c, 2g, and 2k). While comparing the signals II one can easily notice, that this feature is the most pronounced in the case of the sample *C*. This fact is attributed to the contribution of the C=O species from the PCBM molecules as its content was the highest within film *C*. The signals of the higher binding energies are attributed to the C–O and Si–O species.

While analyzing the XPS spectra of the samples B and C in the Si 2p region one observe substrate related peaks that correspond to the Si^0 species (peak II in Fig. III.8.2h and 2l). For the sample B this fact can be explained by the very low thickness of the film (around 5 nm). In sample C the relatively high concentration of the PCBM molecules resulted in the formation of large agglomerates, as revealed by AFM investigations, and therefore the final film was not homogeneously distributed over the substrate. This discontinuity of the material was probably the main reason for the existence of Si^0 signal in the Si 2p core level spectrum of sample C.

Microscopic studies revealed high homogeneities of the films A and B (chapter III. 8.3.2.). Additionally, the thickness of the sample B did not exceed the sampling depth of the XPS system, what is proven by the presence of the substrate related Si^0 feature in the Si 2p core level signal. Therefore, one may suppose that the obtained spectra include the information concerning the chemical composition of the film and the surface of the substrate as well. The XPS quantity analysis [27] of these surfaces has been performed, excluding the substrate information in order to draw conclusions concerning the chemical compositions of the films A and B. The spectra normalization was achieved by dividing the observed relative peak areas by the atomic and instrument sensitivity factors equal to 0.711, 0.296, 0.477, and 0.339 for O, C, N, and Si, respectively [25]. Table 2 lists the atomic ratios of the elements within the analyzed surfaces. The signal coming from the Si^0 feature has not been taken into account.

Table III.8.2. Atomic ratio within the surfaces A and B.

Sample	C	N	O	Si^{4+}
A	4.1	1.1	2.8	1
B	6	1	2.9	1
Unhydrolyzed APTMS monomer	6	1	3	1
poly[(aminopropyl)siloxane] polymer[119]	3	1	1.5	1

Data summarized in the Table III.8.2 indicate, that the APTMS monomers undergo the hydrolysis and condensation reactions during the spin-coating and drying process when exposed to ambient moisture, losing $-OCH_3$ and $-CH_3$ species. The chemical composition of the film B is very similar to that of the unhydrolyzed APTMS monomer, however also in this situation the condensation reaction took place since the material hardened in ambient conditions. The increase of the overall amount of carbon and oxygen species within the sample *B* in comparison to the sample *A* is a result of the PCBM molecules contribution.

III.8.3.1.2. POSS dopants within the APTMS films

In this chapter the influence of the POSS dopants on the chemical properties of the APTMS films is discussed. Samples *D* (Fig. III.8.3a-d), *E* (Fig. III.8.3e-h) and *F* (Fig.III.8.3i-l) have been measured. The sample *D* was investigated after its storage in ambient conditions

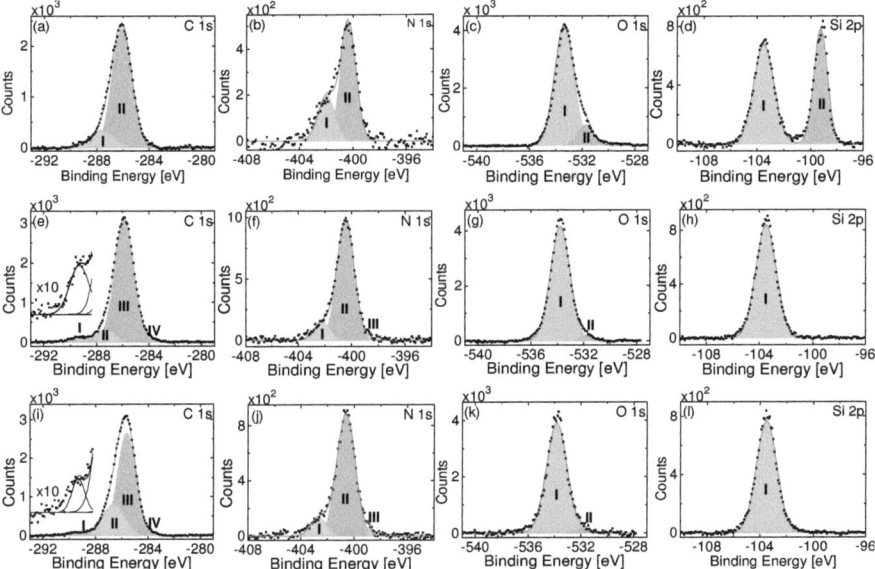

Fig. III.8.3. XPS spectra of the samples *D*, *E* and *F*. Top row (a-d) represents the sample *D*, middle row (e-h) depicts sample *E*, bottom row (i-l) sample *F* respectively. The following core level spectra are drawn in the same order for every sample: C 1s, N 1s, O 1s, Si 2p. The insets in plot (i) and (e) show the magnified signals I.

for one week, while samples *B* and *C* were measured 24 h after the film deposition and storage in ambient conditions. The detailed peak analysis is given in Table III.8.3.

Table III.8.3. Peak decomposition of the XPS spectra of samples *D*, *E*, and *F*.

Region	Peak number	Position [eV]			Assignment		
		D	*E*	*F*	*D*	*E*	*F*
C 1s	I	-------	-289.3	-289.3	--------	NHCOO⁻ [119]	NHCOO⁻ [119]
	II	-287.6	-287.3	-286.6	C–N sp3 [121] &C–O[63]	C–N sp3 [121] &C–O[63]	C–N[63]&C–O[199]
	III	-286.1	-285.9	-285.6	– (CH$_2$)$_3$–[119]	– (CH$_2$)$_3$– [119]	– (CH$_2$)$_3$–[119] &C=C[12]
	IV	------	-283.7	-283.7	-------	C–Si[124]	C–Si[124]
N 1s	I	-402	-402.2	-402.6	–NH^{3+}[119]	–NH^{3+}[119]	–NH^{3+}[119]
	II	-400.4	-400.5	-400.6	–NH$_2$[119]	–NH$_2$[119]	–NH$_2$[119]
	III	------	-398.7	-398.9	--------	C–NH–C[127]	C–NH–C[127]
O 1s	I	-533.3	-533.8	-533.8	–C–O [127]	–C–O[127]	–C–O[127]
	II	-531.7	-531.8	-531.8	C=O& C–NOO[127]	C=O& C–NOO[127]	C=O& C–NOO[127]
Si 2p	I	-103.5	-103.5	-103.5	O–Si–O/Si–OH[128]	O–Si–O/Si–OH[128]	O–Si–O/Si–OH[128]
	II	-99.5	---	---	Si^0[197]	---	---

Again the peak decomposition of the C 1s core level spectra (Fig. III.8.3a, 3e, and 3i) is discussed. Due to the large structures formed as a result of the POSS condensation and aggregation on the surface of sample *D* it was quite difficult to notice a C 1s XPS (Fig. III.8.3a) signal corresponding to the carbamate species of the APTMS. However, the C 1s spectra have been decomposed into two Gaussian peaks. The higher energy feature (II) consists of the C–N and C–O species and the lower energy signal (III) corresponds to the carbon atoms building the POSS functional groups (mainly cyclopentyl) and to the propyl groups of the APTMS.

The C 1s spectra of the samples *E* and *F* have been split into four peaks centered at different binding energies (Fig. III.8.3e, 3i). Contrary to the sample *D,* for these two samples one can easily observe signals corresponding to the carbamate groups at around -298.3 eV (peaks I). The sample *F* contains only 0.5 % of PCBM molecules. Hence, similarly to sample *B* (compare to Fig. III.8.2e), it is difficult to define well pronounced fullerene-cage shake-up satellites in the decomposition process. However, the peak II of the C 1s core level of sample *F* is significantly higher in comparison to sample *E* indicating the presence of oxidized carbon species of the PCBM molecules within the film *F*. An observable shift of the signals III towards lower binding energy in the data of the samples *E* and *F* in comparison to the surface of the pure APTMS is probably caused by the presence of carbon originating from the fullerene cages of PCBM species. The features IV found in the C 1s spectra of these samples are assigned to the C–Si species of the POSS molecules.

Next the N 1s core level peak decomposition of the samples *D*, *E*, and *F* is described. The peaks I and II in the N 1s signals of these samples are assigned to the $-NH^{3+}$ and $-NH_2$ groups respectively. Furthermore, very interesting is an appearance of a weak feature in the N 1s signals of the samples *E* and *F* that are attributed to the C–NH–C species (signal III, [127]). One should to point out that this feature exists in both samples *E* and *F*, where sample *E* contains no PCBM molecules. Therefore, one may explain its appearance by interactions between the POSS and APTMS molecules presumably via a nucleophilic attack of the amine groups of APTMS to the vinylic methylene group of POSS [200,201]. This proposed reaction is illustrated in Fig. III.8.4. Since the environment of this reaction was free of external catalyst only some of the vinyl double bonds have been affected by this process, what is proven by the small areas of the signals III corresponding to the C–NH–C species. The N 1s signal of the sample *D* was too noisy to state doubtless the existence of this weak feature.

The O 1s XPS spectra of the samples *D*, *E*, *F* (Fig. III.8.3c, 3g, and 3k) have been decomposed into two features corresponding to the –C–O groups [23] (peaks I) as well as to C=O and C–NOO [23] species (signals II). One can easily observe, that the contribution of the peak II to the O 1s signal of sample *D* is significantly higher than to the signals of the samples *E* and *F*. This is attributed to the longer storage of sample *D* in the ambient and thus to higher exposition to CO_2.

The Si 2p core level spectrum of the sample *D* contains contrary to samples *E* and *F*, where only Si^{4+} contributions are observed, a high and pronounced feature originating from the Si^0 species of the substrate. This might be a proof of the film *D* inhomogeneity. Therefore this sample was not used in the electrical investigations described below.

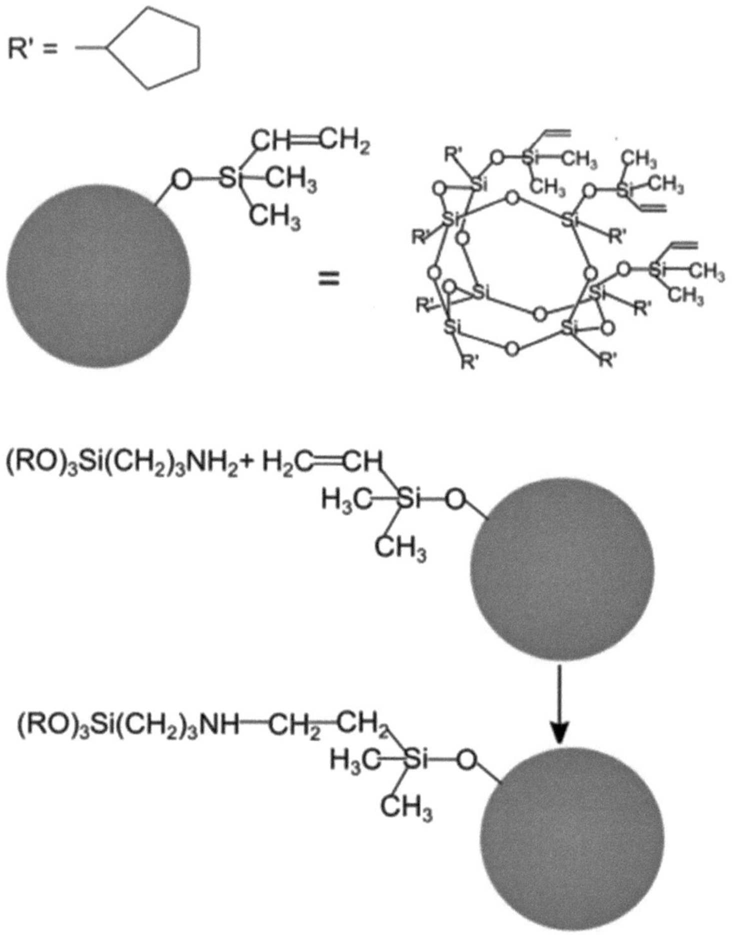

Fig. III.8.4. Scheme of the proposed reaction between APTMS and POSS molecules involving a nucleophilic attack of the primary amine group to a vinylic methylene group. Grey spheres indicate the POSS molecules.

III.8.3.2. AFM investigations

The purpose of using AFM in our investigations was twofold. First, we applied AFM in order to study the surface morphology of the samples *A*, *B*, *D*, *E*, and *F*. Sample *C* was not investigated due to large PCBM aggregates with the mean height of the order of some micrometers. Second, the AFM was used to determine the thickness (*d*) of the deposited materials in order to calculate the dielectric constant from CV measurements (Eq. III.7.3). The thickness of the films was determined by averaging the height values obtained from the surface height histogram analysis performed on different positions on the samples as we described in [202]. The thicknesses of the samples are given in Tab. III. 8. 4.

For morphology investigation at least three images in different positions on the samples were acquired and fluctuations of the root mean square (RMS) surface roughness (*Sq*) of less than ± 6 % were observed. Due to the fact that the APTMS based matrix was used in all investigated films, we used sample *A* as the reference for the surface morphology (Fig. III.8.5a). The calculated *Sq* equal to 0.20 nm is comparable with the result reported in [147] and is similar to the roughness of a Si(001) surface covered with native oxide [148,149]. The surface roughness is very homogenous what is also confirmed by observing the profile roughness *Rq* which is close to *Sq* for most of the analyzed line profiles. An example of a typical APTMS profile with *Rq* equal to 0.21 nm is shown below the 3D image topography in Fig. III.8.5a. In general, the addition of dopants to the APTMS matrix radically altered the surface morphology, e.g. 0.2 % PCBM increases the surface roughness to 0.63 nm (Fig. III.8.5b, sample *B*). Small aggregates exhibiting a height of about 3 nm are present on the surface. A typical profile with two aggregates within a length of 1μm reveals a *Rq* value of 0.80 nm as shown in the lower part of Fig. III.8.5b.

In order to obtain information about spatial properties of the surface we calculated the height-height correlation function (HHCF) which is one of the most commonly used members of the correlation functions [203]. For a surface height profile $h(x,t)$ the HHCF is defined as $H(r,t) = \langle [h(x+r,t) - h(x,t)]^2 \rangle$, where the notation $\langle \cdots \rangle$ symbolizes the averaging over the whole profile, *h* indicates the surface height with respect to the substrate at a position *x* at the time *t*, *r* is the lateral separation length between the two surface heights. For an isotropic surface the HHCF is not dependent on the specific orientation and a new variable $r = |r|$ can be introduced to finally express the correlation function as $H(r,t)$ [203]. Fig. III.8.6 shows a typical example of the HHCF calculated from a profile that was randomly selected on both samples *A* and *B*. The shapes of the HHCF curves reveal typical behavior for self-affine fractal surfaces [203]. From the analysis of the HHCF plots two spatial parameters may be calculated: the correlation length (ξ), and the wavelength (λ). The lateral correlation length is defined as the length beyond which the surface heights are not considerably correlated. Hence, for a surface dominated by islands the lateral correlation length indicates the lateral size of the islands. Additionally, in the region beyond the lateral correlation length ($r > \xi$) the HHCF function for surfaces with islands is oscillatory. The period of the oscillations is characterized by the wavelength corresponding to the average distance between the islands. The two spatial parameters ξ and λ together with Sq deliver precise information about the surface texture.

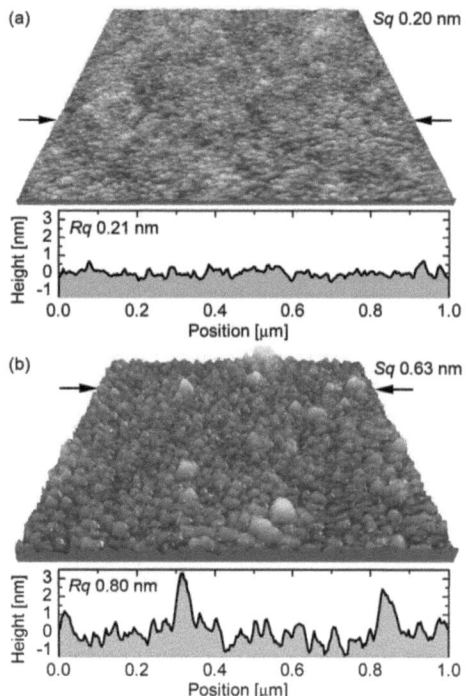

Fig. III.8.5. NC-AFM surface topography of: (a) sample A, color scale 2 nm, (b) sample B, color scale 8 nm. Each image size is 1 μm × 1 μm. Below the 3D AFM images a typical profile acquired at positions indicated by the arrows are shown. The values of the RMS surface roughness Sq and of the RMS profile roughness Rq are also displayed.

We calculated the HHCF at ten different positions on the samples in order to apply statistics. We estimate the correlation length and wavelength deviations in the order of ± 7 % and ±

8 %, respectively. The correlation length equals 15 nm and 19 nm for the samples A (ξ_A) and B (ξ_B), respectively. The wavelength equals 35 nm and 44 nm for the sample A (λ_A) and B (λ_B), respectively (refer to Fig. III.8.6). From these results it is obvious that a small amount of PCBM in a thin APTMS matrix increases not only the surface roughness but also the correlation length and the surface wavelength. We conclude that the AFM method is a valuable supplement to the XPS technique for which the unambiguous verification of PCBM within the sample B was difficult.

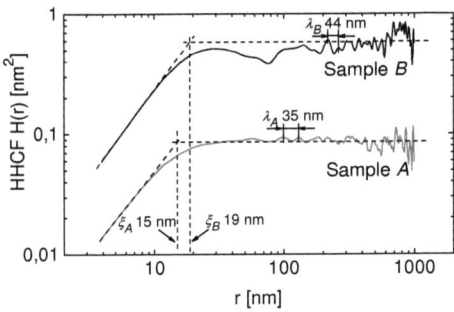

Fig. III.8.6. HHCF calculated for sample A and B. The correlation length ξ and the wavelength λ acquisition positions are marked with arrows.

In the next step of the experiment we investigated the influence of the concentration of the POSS cages in the APTMS matrix on the surface morphology (Fig. III.8.7). In all the samples (D, E, F) the surface topography reveal the existence of structures stimulated by the POSS addition to the APTMS. The process of the inter-molecular POSS-POSS interactions within a continuous medium is explained either by a physical aggregation via a phase separation, or by a chemical POSS-POSS condensation reaction [194]. Since in our preparation routine neither external catalyst nor solvent has been added we presume that the phenomenon responsible for the formation of the observed clusters is dominated by the physical aggregation. In the absence of a stabilizer the overloading of the POSS molecules within the compositions favor the heterogeneous phase separation and the formation of micron-sized inclusions [204]. However, basing on the XPS results, a contribution of the chemical reactions to the POSS domains formation process is also feasible. Fig. III.8.8 illustrates one of the possible condensation reactions between the POSS functionalized APTMS molecules after the hydrolysis reaction due to the ambient humidity.

For 1.5 % POSS concentration in APTMS (sample D, Fig. III.8.7a) the surface is covered by aggregates forming circular fractured patterns with typical diameters ranging from 1 μm to 3 μm.

The average height of the aggregates is 10 nm. For lower POSS concentration of 1 % (sample E, Fig. III.8.7b) the POSS structures tend to form chained and filled patterns with typical dimensions of one structure of 1 μm and 20 nm for the diameter and the height, respectively. Within the films containing 0.5 % POSS and 0.5 % PCBM (sample F, Fig. III.8.7c) structures similar to that obtained for sample E with a decrease of the height to 10 nm are observed.

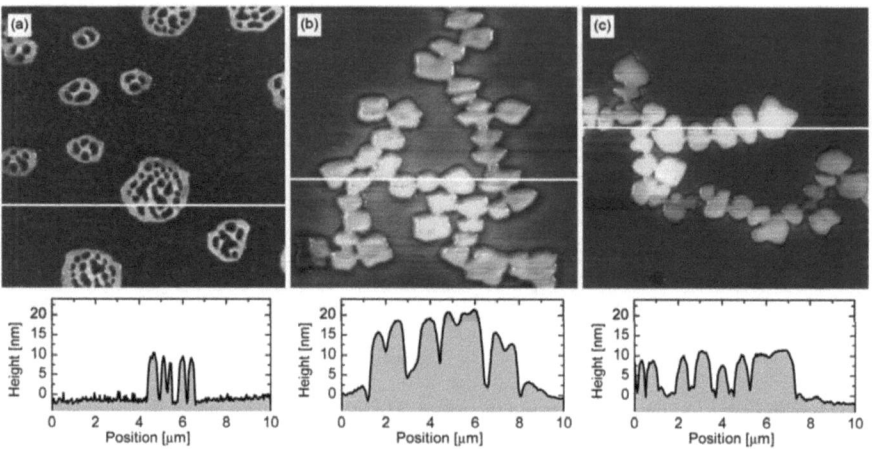

Fig. III.8.7. NC-AFM surface topography and height profile of: (a) sample D, color scale 20 nm, (b) sample E, color scale 35 nm, (c) sample F, color scale 20 nm. Each image size is 10 μm × 10 μm. Horizontal white lines indicate the position where the height profiles were taken.

In Fig. III.8.9. we show 50 μm × 50 μm AFM surface scans of the samples E and F. To quantitatively describe the obtained results we used fractal geometry as this approach is extremely flexible in describing the variety of irregular objects and has been successfully applied for studies of polymer molecules, Earth's continents coastlines and galaxies in the Universe [205]. The shapes of the bright structures which appear on the surface are similar to that described by fractal diffusion-limited aggregates [206] and cluster-cluster aggregation models [207]. We processed the AFM images (Fig. III.8.9a and 9b) by a threshold function in order to expose only the aggregates (Fig. III.8.9c and 9d). It is evident that the sample E (Fig. III.8.9a and 9c) reveals a more compact network of aggregates in comparison to sample F (Fig. III.8.9b and 9d) what is explained by the difference in the POSS concentration between the two samples. To quantitatively compare these two samples we calculated the fractal dimension, which is an equivalent of the aggregate complexity, by using the box-counting

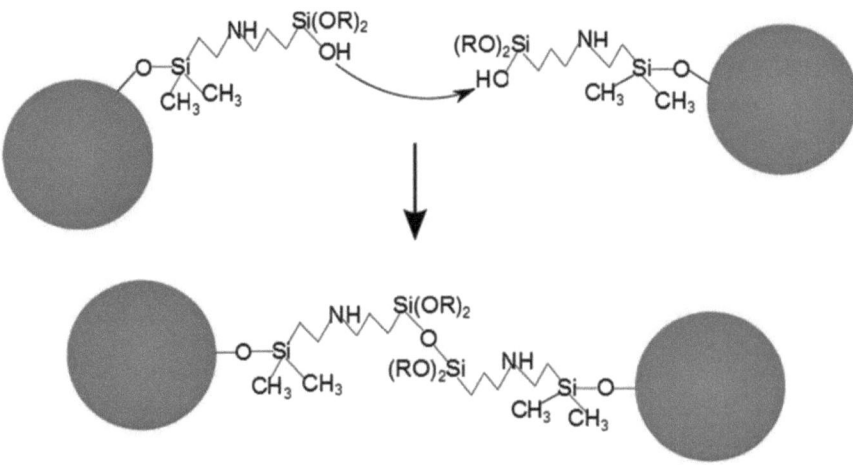

Fig. III.8.8. Scheme of the possible condensation reaction between POSS functionalized APTMS molecules in the presence of ambient humidity. The upper species are formed according to the reaction presented in Fig. III.8.4, where the –OH groups are a product of the –OCH$_3$ groups hydrolysis due to the ambient humidity.

method [208]. The obtained fractal dimension equals 1.11 and 1.29 for sample E and F, respectively, confirming that a higher POSS concentration within the APTMS matrix promotes the appearance of more complex aggregates on the surface. Additionally, on both samples also pits with mean depth of 15 nm are found.

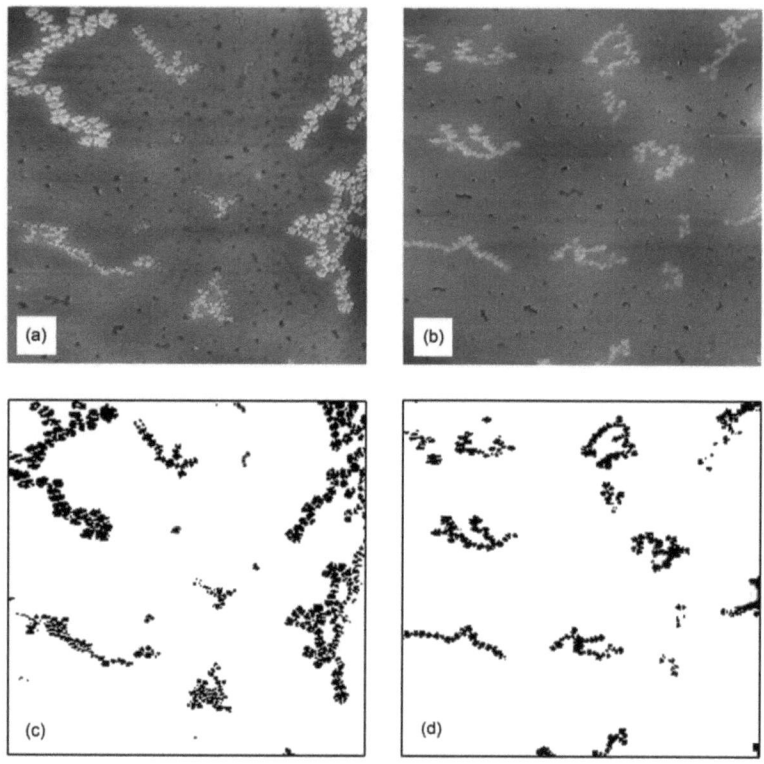

Fig. III.8.9. NC-AFM surface topography of: (a) sample E, and (b) sample F. The color scale is the same for the topography images and equals 60 nm. The corresponding processed images show only aggregates of POSS species of: (c) sample E, and (d) sample F. Each image size is 50 μm × 50 μm.

III.8.3.3. CV measurements

Next the CV measurements performed on the samples A, B, E and F are reported. As described above the samples C and D were excluded for these measurements. Also for sample B we do not present the CV results because this layer was very thin. AFM height analysis [202] as well as XPS analysis of the signal attenuation of the substrate [209] delivered a thickness in the range of 5 nm. Due to the small thickness of this layer we observed high leakage currents which ruled out the permittivity calculation via the accumulation capacitance.

CV measurements of the chosen samples were carried out at least at three different contacts distributed over the sample surface, i.e. different contact areas were used. In Fig. III.8.10 the CV characteristics of pure (sample A) as well as of POSS (E) and PCBM/POSS (F) incorporated

APTMS films are shown. In order to compare these data easily, the measured capacitance values (C_{MIS}) were "normalized" to the thickness (d) of the low-k layer and the vacuum permittivity (ε_0) into the form $C_{MIS}*d/\varepsilon_0$. The plotted CV curves in Fig. III.8.10 were recorded at the same contact, i.e. the contact area in all three shown measurements was the same within a deviation of 6.5 %. This deviation was taken into account in Fig. III.8.10 by an area correction factor reflecting the contact area deviation of every sample compared to sample A. In that way the reader can easily compare now the accumulation capacitance of these three samples as an equivalent value of their permittivity (refer to Eq. III.7.3). We observe that the APTMS exhibits the highest permittivity which is lowered by the incorporation of POSS and in particular of combined PCBM and POSS. This will be further analyzed below.

In Fig. III.8.10 we further realize distinct hysteresis in all three samples. In chapter IX we have already reported about hysteresis in C_{60} and PCBM incorporated APTMS films. In the results of the present work we observe that the hysteresis is already strongly existent and most prominent in the pure APTMS (sample A) films. The direction of the hysteresis suggests the injections of charge carriers from the metal into trap centers of APTMS [134]. In the literature nitrogen related defects [134] and oxygen vacancies [134] are for example discussed as probable trap centers in APTMS based films. In our XPS C 1s and N 1s core level spectra we found carbamate (NHCOO⁻) and NH_3^+ species, respectively, due to the interaction with ambient CO_2. These species might act as charge trapping centers. As in sample A the intensity of the NH_3^+ related signal in the N 1s spectra is much stronger compared to the signal attributed to the carbamate in the C 1s signal, we argue that the mechanism of a filling and defilling of NH_3^+ proposed by Chauhan et al. [134] is most probable in our APTMS sample. Furthermore, we have to note that we have not yet performed annealing steps during the preparation of our layers. Therefore also moveable charges might correspond to the hysteresis within the CV loops. Si–OH groups or water adsorbed at the Si surface might be responsible for that [118,210]. Our APTMS layer was not fully hydrolyzed as reported in the XPS result section, therefore –OH groups and water might still be present at the Si surface limiting the silane group reaction on the substrate surface [210].

In the POSS and PCBM/POSS incorporated films we found lower hysteresis, in particular in the PCBM/POSS sample. Comparing the NH_3^+ contributions in the N 1s core level signals of these layers with that one of the pure APTMS we observe reduced signals for the samples E and F. The contribution of the NH_3^+ species to the signals I in the N 1s core level equals 13 %, 12 % and 11.4 % for the samples A, E, F respectively. This fact supports the idea of the filling/defilling of the NH_3^+ traps discussed above. In addition we observe a difference of the middle position of the CV curves (i.e. zero flat band voltage without mobile charges or filled/defilled traps) in respect to the voltage axis in particular between the samples *A*, *E* and the sample *F*. Assuming a vacuum work function of silver of 4.1-4.3 eV [38,211,212], using an electron affinity in Si of 4.05 eV [38], and calculating a mean difference between the conduction band minimum and the Fermi level of Si related to the doping level of the Si wafer [38] in the range of 0.25 eV, a work function difference between the silver metal and the Si substrate in the range of approximately -0.2 eV to 0 eV can be proposed. In Fig. III.8.10 we notice that the flat band voltage of sample *F* is within that range, whereas those of the samples *A* and *E* are positively shifted along the voltage axis. Besides possible interactions between the metal or the substrate with the low-*k* layer this might be due to negative fixed charges within the low-*k* layer. Indeed, we realize higher contributions of the NHCOO$^-$ species in the N1s core levels of the samples *A* and *E* in comparison to sample *F*.

Now we return to the determination of the relative dielectric constant (*k*) of these films. Therefore we took the normalized capacitance values like depicted in Fig. III.8.10 (without the described area correction factor) in the accumulation regime (C_{acc}) and plotted these values in Fig. III.8.11 versus the contact area (*A*). Converting the plate capacitor equation into Eq. III.7.3 it can easily be seen, that corresponding to Eq. III.7.3 the slope of a linear fit to the data delivers the *k*-value of the films. In Fig. III.8.11 the data of the sample of 1 % PCBM in APTMS (sample *G*) reported in chapter III.7. are also shown for comparison. The details of the samples and the fit results are listed in Table III.8.4. Please note that possible SiO_2 formations or residual SiO_2 at the Si surface has not been taken into account during the permittivity determination as the investigated films are relatively thick and therefore the influence of a probable additional interfacial capacitance can be neglected (chapter III.7.). Within this comparison we observe that by mixing PCBM into APTMS (sample *G*) the permittivity of the film can be reduced. By adding a co-mixture of PCBM and POSS into APTMS (sample *F*) this value can be further decreased, whereas the POSS incorporation into APTMS (sample *E*) itself shows no distinct improvement compared to pure APTMS. The determined dielectric constant of APTMS is relatively high compared to other literature values of APTMS based films [178,213]. However, we have to state again that we have not yet performed any heat treatment. As discussed above due to the not complete hydrolysis reaction and to the interactions with ambient CO_2 residual groups like –OH, NH_3^+, and also water may remain at the surface or within the layer [134,210]. The high polarizabilities of these species may strongly affect the overall polarizability and hence will increase the permittivity of the layer [118,134,214]. Nevertheless our CV investigations indicate that the overall permittivity of the APTMS layer can be reduced by co-adding PCBM and POSS dopants into the polymer leading to hybrid nanocomposites with improved properties.

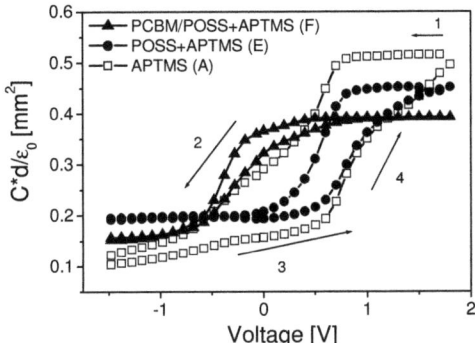

Fig. III.8.10. CV characteristics of samples *A* (open squares), *E* (filled circles) and *F* (filled triangles). The capacitance values are depicted in a normalized way as described in the text. The measurement direction is indicated by the indexed arrows.

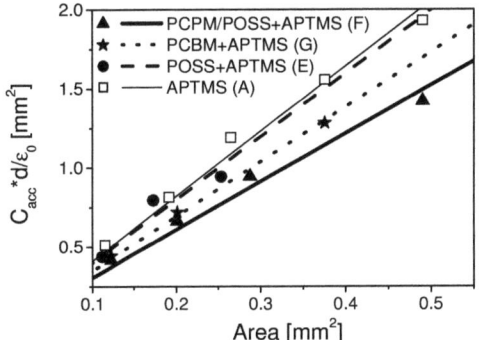

Fig. III.8.11. Normalized accumulation capacitance versus the contact area of samples *A* (open squares), *E* (filled circles) and *F* (filled triangles). For comparison a sample consisting of 1 % PCBM in APTMS (filled stars, sample G) taken from [202] is shown too. The normalization is performed regarding Equ. III. 7.3, which was also used in order to apply linear fits to the data (sample *A*: thin solid line, *E*: dashed line, *F*: thick solid line, *G*: dotted line). The results of the fits are listed in Table III.8.4.

Table III.8.4. Initial compositions and thicknesses of the electrically investigated samples and relative dielectric constant deduced by fits according to Eq. III. 7. 3.

Sample	APTMS [%]	PCBM [%]	POSS [%]	Thickness [nm]	ε_r
A	100	0	0	138	4.1
E	99	0	1	219	4.0
F	99	0.5	0.5	185	3.0
G	99	1	0	57	3.5

III.8.4. Conclusions

The main goals of the described in this chapter experiment were the investigation of the interactions between PCBM, POSS and APTMS within the produced hybrid low-*k* material and the estimation of the influence of the PCBM and POSS molecules on its properties. In order to realize this purpose the investigations by means of spectroscopic, microscopic and CV techniques have been performed. The combined application of these methods opens the opportunity of gathering complementary information concerning chemical and electrical properties of the analyzed material and its surface morphology as well.

The XPS spectra of the analyzed samples indicate that the POSS molecules more likely undergo a chemical reaction with the APTMS than with the PCBM. Those interactions proceed presumably via a nucleophilic attack of the amine groups of APTMS to the –C=C– groups of POSS. The characteristic shake-up features of the fullerene cages that are present in the C 1s core level XPS spectra of the PCBM containing samples suggest, that a relatively large amount of –C=C– species building those molecules remained intact. However, for the layer of very low PCBM concentration (0.2 %, sample *B*) an unambiguous verification of PCBM is difficult. In this situation the AFM method is a valuable supplement to the XPS technique since even a small amount of PCBM in a thin APTMS matrix increases not only the surface roughness but also the correlation length and surface wavelength, as revealed by our investigations. The XPS investigations reveal furthermore a significant influence of the ambient humidity onto the chemical properties of the produced films. Due to the adsorbed H_2O molecules a series of chemical reactions within the deposited films take place, including hydrolysis, condensation and nucleophilic attack as well. The influence of the ambient CO_2 resulting, among others, in the formation of carbamate species, is also significant. The substrate Si^0 features in the XPS spectra of the samples C and D are due to a large agglomerate formation within the layers as revealed by the AFM investigations (not shown for the sample C). Those cluster formation resulted in the discontinuity of material distribution on the substrate prohibiting a detailed CV characterization. The film of the sample B is homogenous and continuous, thus the existence of the Si^0 feature in the Si 2p core level is related to the relatively low thickness of the deposited material (around 5 nm). Summarizing the XPS data of all samples (*A-F*) one may draw the conclusion that the information obtained by the spectroscopic and microscopic studies are complementary.

AFM investigations revealed, that the amount of the POSS monomers within the produced layers was overloaded resulting in a fractal-shaped cluster formation due to the aggregation process. In order to obtain well dispersed films the application of a significantly lower POSS concentration in the presence of a stabilizer is needed. In particular this improvement in the composition properties can often be found even at a relatively low filler content [190].

CV measurements pointed out that the dielectric constant of the pure APTMS films obtained in our experimental conditions equals 4.1 and is not significantly influenced by doping it with the pure POSS molecules, while at the same time the application of a mixture of PCBM and POSS cages as nanodopants decreases this value to 3.0. The previous experiment (chapter IX) showed that the application of pure PCBM molecules decreases the APTMS dielectric constant to 3.5. One should note, that the obtained values might be overestimated due to the presence of the species possessing a high polarizability, as confirmed by the XPS analysis (–OH, NH_3^+, $NHCOO^-$, and H_2O). The hysteresis in the CV measurements was attributed to the NH^{3+} species within the films also confirmed by XPS results.

In order to realize organo-funcionalized siloxane networks with incorporated fullerene cages and POSS molecules an external catalyst would be required for the completion of the addition, hydrolysis and condensation reactions within the composite. However, at the present stage of investigations this work demonstrates that by the co-adding of PCBM and POSS dopants into the APTMS matrix, hybrid nanocomposites with improved properties compared to the pure matrix can be achieved.

III.9. Spectroscopic and capacitance-voltage characterization of thin aminopropyl-methoxysilane films doped with copper phthalocyanine, tris(dimethylvinylsilyloxy)-POSS and fullerene cages[7]

III.9.1. Introduction

Phthalocyanines are conjugated molecules of outstanding optical and electrical properties with high thermal and chemical stability. The wide range of possible applications of this large family of compounds includes electrochemical devices, solar cells, gas sensors, optical switches, data-storage media and many others [62]. Metallophthalocyanianes are considered for various technological applications like organic transistors, chemical sensors, and organic electroluminescent devices [58]. Thin organic films composed of MtPc such as copper phthalocyanine (CuPc) attracted a great scientific attention in order to design potential molecular devices [55]. The literature reports also about the application of CuPc oligomers as high dielectric constant fillers in a polymer matrix [58,59]. According to Zhang et al. the observed dielectric constant of CuPc oligomers has been as high as 10^5 [58]. However, the work using the theory of the local dielectric permittivity performed by Shi and Ramprasada [59] indicates, that CuPc monomers do not intrinsically have a high dielectric constant, as values of arround 15 and 3.4 along and perpendicular, respectively, to the CuPs plane were found. These authors attributed the high value of the CuPc systems to some other extrinsic factors [59]. In our experiment CuPc has been applied as a filler into low-k aminopropylsiloxane based thin films. As revealed by our CV investigations the presence of the CuPc filler within the APTMS based thin films decreases the value of the effective dielectric constant of the resulting matrix.

As an additional filler to some of the low-k composites described in this chapter [6,6]-phenyl-C_{61}-butyric acid methyl ester (PCBM) has been applied. PCBM has been used as a source of fullerene cages that are known for its unique physicochemical properties [60]. In this chapter PCBM has been incorporated into a siloxane matrix in order to decrease the dielectric constant of the resulting material by increasing the porosity within the films [61]. Since the non-functionalized C_{60} fullerene exhibits a low solubility and the tendency to form agglomerates in solvents [107,191], the utilization of PCBM as its better soluble derivative allows to obtain a better dispersion of the fullerene species within the films.

---

[7] J. Klocek, K. Henkel, K. Kolanek, E. Zschech, D. Schmeißer, Spectroscopic and capacitance-voltage characterization of thin aminopropylmethoxysilane films doped with copper phthalocyanine, tris(dimethylvinylsilyloxy)-POSS and fullerene cages, Applied Surface Science, 258 (2012) 4213-4221

In order to further decrease the dielectric constant by introducing porosity on the one hand [57,61] and to increase the oxidation resistance on the other hand some films have been additionally reinforced with incompletely condensed frameworks possessing a hybrid inorganic-organic, three-dimensional structure cage polyhedral oligomeric silsesquioxanes (POSS). The incorporation of this kind of molecules into the polymeric material allows to control the dielectric constant. Furthermore, this can lead to many substantial varieties and improvements in final compositions ranging from the thermal stability, hydrophobicity, refractive index, oxidation resistance, surface hardening to the optical clarity, what attracted a great research interest in the past few years [57,193–195].

The chapter is organized as follows: Based on the results of X-Ray photoelectron spectroscopy (XPS) the chemical composition and the resistance against ambient influences of the hybrid material films composed of the CuPc, POSS and PCBM species incorporated in various concentrations into the APTMS based siloxane matrix is discussed. The next section describes the electrical characterization of the samples by the capacitance-voltage (CV) method. Finally, a summary of the relevant conclusions is given.

III.9.2. Experimental details

III.9.2.1. Materials and substrates

In this work PCBM (99.5 %), APTMS (97 %) and tris(dimethylvinylsilyloxy)-POSS (further referred as POSS in the text), obtained from Sigma-Aldrich have been used. CuPc (95 %) has been delivered from Alfa Aesar. Phosphorous doped n-type silicon (100) wafers having a resistivity of 1-5 Ωcm delivered from Crystec (Berlin, Germany) were chosen as substrates for our samples.

III.9.2.2. Sample preparation

The schemes of the investigated samples are illustrated in Fig. III.9.1 (samples A, B, C with higher dopant concentrations) and Fig. III.9.5 (samples D, E, F with lower dopant concentration). The dopant concentrations within the films were calculated in weight percent.

After 10 minutes of ultrasonification in isopropanol and then in distilled water, the Si substrates were dried in nitrogen stream. The wafer surfaces were hydroxylated by the immersion in piranha solution (a mixture of 7 : 3 (v/v) 98 % H_2SO_4 and 30 % H_2O_2) at 120°C for 1.5 h. The substrates were then rinsed with distilled water and isopropanol several times and were dried afterwards in nitrogen stream.

The hybrid materials were produced from a dispersion of PCBM, CuPc and POSS particles in the APTMS monomer. The thin films were produced by spin-coating at a speed of 6000 rpm for 15 seconds in glove box conditions under argon atmosphere.

Additionally, CV investigations were performed on the samples *A, B, D,* and *F* as these samples had very uniform thickness distributions. On top of these samples shown in Fig. III.9.1 and 4 silver contacts with diameters between 400 µm and 800 µm were evaporated through a shadow mask in order to finalize the MIS structure of the samples. Results achieved previously on a pure APTMS sample (chapter III.8.) are appended for comparison in the CV analysis section of this chapter.

III.9.3. Results and discussion
III.9.3.1. XPS investigations on the samples with higher CuPc concentration (>=0.4 %)

In Fig.III.9.1 the schemes of the samples described in this part are presented. The sample *C* has been additionally reinforced with POSS molecules. Corrections for the energy shift were accomplished assuming a binding energy of 103.5 eV for the Si 2p peak of the Si^{+4} species [119]. Fig. III.9.2 represents the XPS results obtained for the samples *A, B* and *C*. The detailed peak analysis is listed in Table III.9.1.

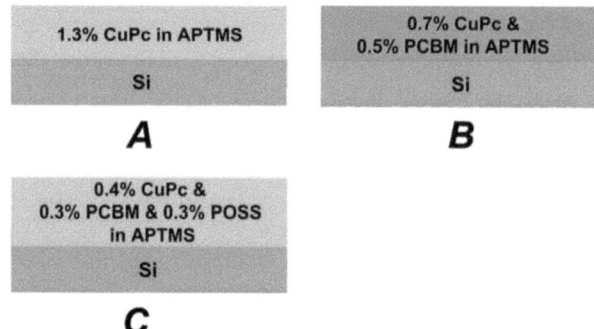

Fig. III.9.1. The schemes of the investigated samples containing higher dopant concentration prepared by spin coating onto the Si substrate: *A* 1.3 % CuPc solution in APTMS (referred in the text as sample *A*), *B* 0.7 % CuPc and 0.5 % PCBM solution in APTMS (sample *B*), *C* 0.4 % CuPc, 0.3 % PCBM and 0.3 % POSS solution in APTMS (sample *C*).

The C 1s XPS spectra of the samples *B* and *C* contain the characteristic π-π* shake-up peaks which confirm the presence of fullerene cages on the surface (peaks I'). These features are absent in the data of the sample *A* where the APTMS matrix were doped only with the CuPc molecules. Nevertheless, the presence of the –C=C– species coming from both CuPc and PCBM molecules may also be confirmed during the observation of the shift of the main C 1s signals (peaks III). In general, these features represent simultaneously existing sp^2 (graphite like) and sp^3 (diamond like) carbon contributions. Since APTMS is the basic component of the analyzed films one may presume, that the main C 1s signal of the XPS spectra is dominated by propyl groups of this substance which signal contribution arises at a center position around 286 eV [119,123]. Thus, a shift of the center position of this feature towards lower binding energy is an evidence of the increase of the sp^2 contribution to the C 1s

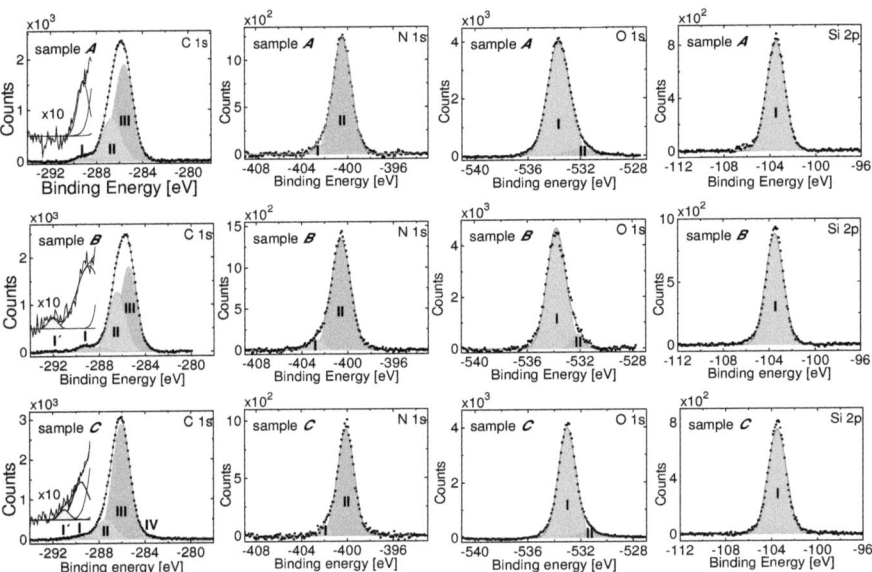

Fig. III.9.2. XPS spectra of the samples A (top row), B (middle) and C (bottom row). The following core level spectra are drawn in the same order for every sample: C 1s, N 1s, O 1s, Si 2p. The insets within the C1s spectra show the magnified signals I´.

core-level spectra. The main feature in the C 1s spectra of pure fullerene and CuPc films appears around 284.6 eV; 284.7 eV [63,65] suggesting a large contribution of the sp^2 species within the films composed of these substances. The center positions of the main feature within the C 1s data of the samples A and B are very similar (285.6 eV and 285.5 eV respectively). These binding energies are lower than those corresponding to the C 1s main feature obtained for the pure APTMS films [119,123]. It suggests, that the content of dopants including sp^2 carbon species (CuPc, PCBM) was high enough to shift the center position of the main feature towards lower binding energy. In contrast, the center position of the main C 1s feature of the sample *C* is in a good agreement with the literature data concerning pure APTMS (286.1 eV). Nevertheless, the fitting procedure delivered an additional feature at around 284 eV (peak IV) within the C 1s spectrum of this sample. This signal is assigned to a hybrid feature originating to Si-C species of the POSS molecules and sp^2 carbon contributions from CuPc and PCBM.

Table III.9.1. Peak decomposition of the XPS spectra of the samples *A*, *B*, and *C*.

Region	Peak number	Position			Assignment		
		A	B	C	A	B	C
C 1s	I´	---	-292.0	-291.0	---	π-π* shake-up [138]	π-π* shake-up[120]
	I	-289.2	-288.9	-289.6	NHCOO$^-$ [119]	NHCOO$^-$ [119]& π-π* shake-up[120]& –O–C=O[67]	NHCOO$^-$ [119]
	II	-286.4	-286.4	-287.4	C–N[63]& C–O[199]	C–N[63]& C–O[199]	C–N sp^3 [121]&C–O[63]
	III	-285.6	-285.5	-286.1	–CH$_2$– [119] &C=C[12]	–CH$_2$– [119] &C=C[12]	–CH$_2$– [119]
	IV			-284			C-Si[124]
N 1s	I	-402.5	-402.8	-401.9	NH$_3^+$[119]	NH$_3^+$[119]	NH$_3^+$[119]& NHCOO$^-$[119]
	II	-400.5	-400.6	-400.1	–NH$_2$[119]	–NH$_2$[119]	–NH$_2$[119]
O 1s	I	-533.7	-533.8	-533.0	C–O–C,C–O–O$^-$,C–OH[67]	C–O–C,C–O–O$^-$,C–OH[67]	C–O–C,C–O–O$^-$,C–OH[67]
	II	-531.9	-532.2	-531.2	C=O, O–C=O[67]	C=O, O–C=O[67]	–C=O[126]& -N–C=O[127]
Si 2p	I	-103.5	-103.5	-103.5	O–Si–O/Si–OH[128]	O–Si–O/Si–OH[128]	O–Si–O/Si–OH[128]

While comparing the C 1s core level spectra of the discussed samples one notice, that the feature II is the smallest in the data of sample *C* where POSS molecules have been incorporated into the film. Taking into account that a significant contribution to these signals has been attributed to the carbon oxidized states one may suppose that the surface of the film doped with POSS molecules (sample *C*) was the most resistant against oxidation in ambient conditions.

The conclusion regarding the higher resistant of the POSS containing sample against the interactions with ambient components can be confirmed while analyzing decomposed XPS spectra in the N 1s core level. The area of the feature I corresponding to the NH_3^+ species is the smallest for sample *C*. This protonated amino group is a product of the reactions III.5.1 and III. 5. 1 between APTMS and ambient CO_2 [119]. Thus one may presume, that the smallest area of the corresponding N 1s feature I in sample *C* may be a hint that this surface was the least affected by the ambient.

Quantitative peak analysis [27] was carried out in order to determine the surface element concentrations. The spectra normalization was achieved by dividing the observed relative peak areas by the atomic and instrument sensitivity factors equal to 0.711, 0.296, 0.477, and 0.339 for O, C, N, and Si, respectively [25]. The resulting surface atomic concentrations of the films *A*, *B* and *C* are given in the Table III.9.2.

Table III.9.2. Atomic concentrations of the samples *A*, *B*, and *C*.

sample	% C	% N	% O	% Si
A	45.8	13.1	29.5	11.6
B	45.5	14.5	28.5	11.5
C	56.2	8.9	24.2	10.7

The stoichiometry of an unhydrolyzed APTMS monomer (excluding hydrogen atoms) corresponds to C_6NSiO_3, whereas the "ideal" poly[(aminopropyl)siloxane] polymer has a composition of $C_3NSiO_{1.5}$ [119]. Thus, the content of carbon and the oxygen within the pure APTMS film depends on the degree of the monomer condensation that takes place due to the humidity in ambient conditions. Fig. III.9.3 describes the possible hydrolysis (a) and condensation (b) reactions of the APTMS monomer due to the ambient water.

Although the reactions presented in Fig. III.9.3 lead to changes of the C and O amount within the APTMS based film this process should not influence the Si: N ratio that equals 1.

While analyzing the quantitative surface compositions given in Table III.9.2 one notice very similar compositions for the samples *A* and *B*. Surprisingly, the surface oxygen concentration of the sample *A* is slightly higher than this one of the sample *B*. However, contrary to the sample *B*, sample *A* doesn't consist of oxygen containing PCBM molecules. Thus, one may attribute the highest amount of oxygen species within surface *A* to its relatively strong oxidation in ambient conditions. Taking into account that film *A* contains the highest CuPc concentration of all samples described in this chapter one may suppose, that this dopant accelerates the surface oxidation process.

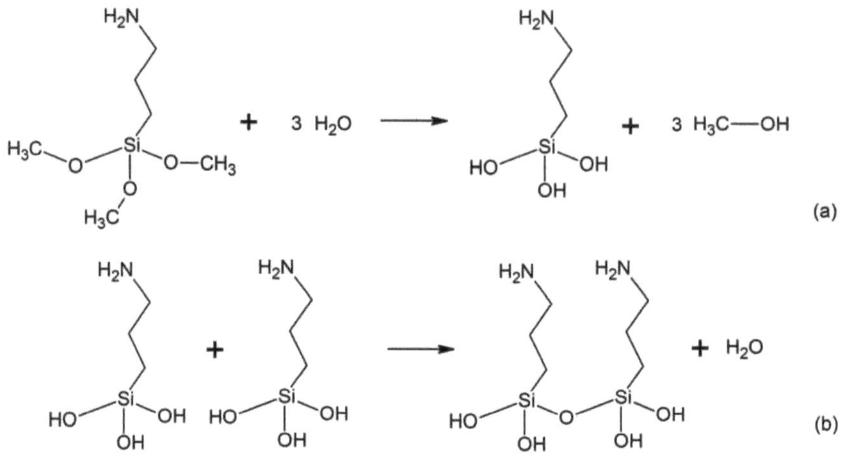

Fig. III.9.3. Scheme of the hydrolysis (a) and the condensation (b) reactions of the APTMS molecule.

The surface elemental composition of sample C suggests, that the POSS molecules incorporation increased the surface carbon concentration to 56 %. Taking into account that this series of samples was produced and stored under the same conditions we do not attribute this fact distinctly to the lower degree of the APTMS monomer condensation. In particular this argument is supported by the lower oxygen content within this surface compared to the samples A and B what on the other hand may also suggest the lower degree of the surface oxidation of sample C. One may observe further that a relatively small amount of POSS molecules incorporated into the film C (0.3 %) are sufficient to increase the atomic Si: N ratio, which is about 1.20 in sample C whereas it is only 0.79 to 0.89 for sample B and A, respectively. Additionally one may conclude that the surface C is the most resistant against oxidation since the detected surface oxygen concentration is lower than for the samples A and B although POSS cages are a source of oxygen (Fig. III.9.4).

III.9.3.2. XPS investigations on the samples with lower CuPc concentration (<0.1 %)

After the aforementioned investigations of the CuPc, PCBM and POSS molecules influence on the APTMS matrix properties similar APTMS based films but with lower concentrations of CuPc and POSS dopants have been produced. On the other hand in this part

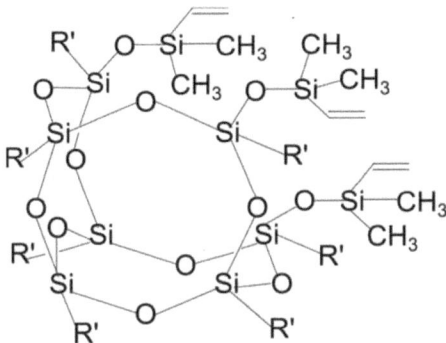

Fig. III.9.4. Scheme of the POSS molecule applied as a dopant into the APTMS matrix (R` are cyclopentyl groups).

of the experiment no oxygen-containing PCBM molecules have been used in order to investigate the influence of the POSS molecules on the surface resistance against oxidation in ambient conditions. The schemes of the samples described in this part are illustrated in Fig. III.9.5; Fig. III.9.6 represents the XPS results obtained for the samples D, E and F. The detailed peak analysis is given in Table III.9.3.

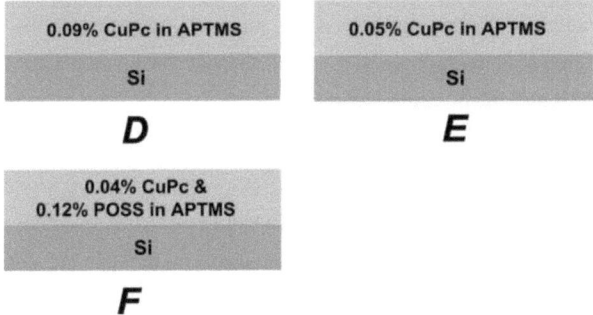

Fig. III.9.5. The schemes of the investigated samples containing lower dopant concentration prepared by spin coating onto the Si substrate: D 0.09 % CuPc solution in APTMS (referred in the text as sample D), E 0.05 % CuPc solution in APTMS (sample E), F 0.04 % and 0.12% POSS solution in APTMS (sample F).

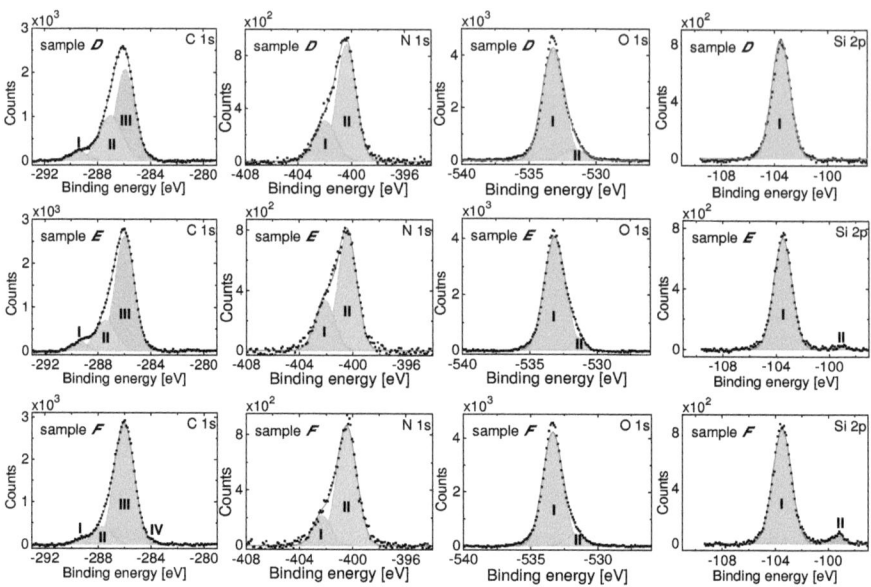

Fig. III.9.6. XPS spectra of the samples *D* (top row), *E* (middle) and *F* (bottom row). The following core level spectra are drawn in the same order for every sample: C 1s, N 1s, O 1s, Si 2p.

The application of dopants influenced the physical properties of the APTMS solutions providing the possibility of the preparation of transparent and homogenous layers in the spin-coating process (thickness in the range of 7-20 nm, see chapter III.9.3.3). Its influence on the chemical properties of the APTMS layers can be estimated with the help of the XPS technique. The centre positions of the features III in all C 1s XPS spectra of the samples discussed in this part are in a good agreement with the literature value of the binding energy corresponding to the propyl group of pure APTMS (286 eV, [119,123]). Because, contrary to the samples *A*, *B*, *C*, no shift of these features towards lower binding energy is observed, we suppose, that either the amount of the sp^2 species has not been detectable due to the low doping concentration or it has been minimized by interactions with the APTMS matrix.

While comparing the features II in the C 1s core levels, attributed to a considerable extent of the carbon oxidized states, one observe also for this sample series that this signal is the weakest for the data of the POSS reinforced film (sample *F*). One may attribute this fact likewise for the thicker film *C* to the smaller surface oxidation in comparison to the other samples. Moreover, also the features I in the C 1s and N 1s of sample *F* corresponding to the $NHCOO^-$ and $-NH_3^+$ species, respectively [119], are clearly smaller. Both species are formed as a result of APTMS interactions with the ambient CO_2 regarding the reactions III.5.1 and III. 5. 1 [119].

Table III.9.3. Peak decomposition of the XPS spectra of samples *D, E,* and *F.*

Region	Peak number	Position			Assignment		
		D	*E*	*F*	*D*	*E*	*F*
C 1s	I	-289.4	-289.3	-289.4	NHCOO⁻[119]	NHCOO⁻[119]	NHCOO⁻[119]
	II	-287.0	-287.4	-287.7	C–N[121]& C–O[63]	C–N[121]& C–O[63]	C–N[121]& C–O[63]
	III	-285.9	-286.0	-286.0	–CH_2–[119]	–CH_2–[119]	–CH_2–[119]
	IV			-283.8			C–Si[124]
N 1s	I	-402.0	-402.1	-402.3	–NH_3^+[119]	–NH_3^+[119]	–NH_3^+[119]
	II	-400.4	-400.4	-400.4	–NH_2[119]	–NH_2[119]	–NH_2[119]
O 1s	I	-533.2	-533.1	-533.4	–C–O[127], –Si–O[128]	–C–O[127], –Si–O[128]	–C–O[127], –Si–O[128]
	II	-531.6	-531.4	-531.6	–N–C=O[127]& C=O[126]	–N–C=O[127]& C=O[126]	–N–C=O[127]& C=O[126]
Si 2p	I	-103.5	-103.5	-103.5	O–Si–O/ Si–OH[128]	O–Si–O/ Si–OH[128]	O–Si–O/ Si–OH[128]
	II		-99.2	-99.3		Si^0[197]	Si^0[197]

The relatively weak signals corresponding to the by-products of the interactions between the film surface and the ambient CO_2 suggest a higher chemical stability of the surface of the sample *F* after air exposure. However, the decrease of the feature II area in the film *F* C 1s core level spectrum in comparison to the corresponding signals of samples *D* and *E* may be also related to some extent to the smallest concentration of CuPc within sample *F*.

The Si 2p core level spectra of the samples *E* and *F* contain the substrate related feature coming from the Si^0 at around 99 eV. This is attributed to the very low thickness of the deposited material (in the range of 10 nm) that did not exceed the sampling depth of our XPS system.

Also for this sample series a quantitative XPS analysis has been performed likewise as done for the samples *A, B,* and *C* reported above. Observing the atomic ratios of the elements within the particular films, the influence of dopants on the chemical composition of the surface can be discussed more distinctly than it would be possible by a qualitative analysis. The surface atomic concentrations of the films *D*, *E*, *F* are given in the Table III.9.4.

Table III.9.4. Atomic concentrations of the samples *D*, *E*, and *F*.

sample	% C	% N	% O	% Si
D	48.6	10.8	30.7	9.9
E	52.0	10.2	28.8	9.0
F	50.8	10.1	28.7	10.4

The stoichiometries obtained as results of the quantitative analysis are: $C_{4.91}N_{1.09}SiO_{3.10}$, $C_{5.77}N_{1.13}SiO_{3.20}$ and $C_{4.88}N_{0.96}SiO_{2.76}$ pointing out Si: N ratios of 0.92, 0.88 and 1.04 for the samples *D*, *E*, and *F*, respectively. For the samples *D* and *E* the slight superiority of the N in comparison to the Si is obviously caused by the contribution of the nitrogen species of CuPc, similar as it has been observed for the samples *A* and *B*. However, due to the very low CuPc dopant concentration the difference of the Si: N ratios between the pure APTMS and the analyzed films is small. For the film *F* reinforced additionally by Si containing POSS a slight superiority of Si in comparison to N is observable, however the Si: N ratio is lower than for sample *C* described above where this value equaled around 1.20.

The quite high amount of C atoms within the films *D* and *E* in comparison to the thicker films with even higher doping concentrations aforementioned may be explained by the adsorption and interactions with ambient CO_2, that due to the small thickness of the films could stronger affect the chemical compositions of the layers than in the samples *A* and *B*. A lower degree of the APTMS condensation of the films *D* and *E* resulting in a higher C content cannot be excluded. For the sample *F* we notice a lower atomic C concentration than for the sample *C*. This may be attributed to the lower dopant concentration within the film *F*. Comparing the atomic ratios obtained for the discussed films with those of the unhydrolysed ATPMS monomer and the "ideal" poly[(aminopropyl)siloxane] polymer one may confirm the conclusions drawn above that the POSS reinforced surfaces are more resistive against the ambient oxidation. The oxygen atomic concentration within the film *F* in only slightly lower than that calculated for the sample *E* but at the same time one should emphasize one more time that the POSS molecule contains a considerable amount of oxygen bonded to silicon atoms (Fig. III.9.4). The concentration of POSS cages within the film *F* (0.12 %) turned out to be sufficient to increase the Si: N ratio within the APTMS matrix although this material contains an additional source of nitrogen (0.04 % CuPc). Thus one may conclude that some contribution to the oxygen atomic concentration within the film *F* originates from POSS cages.

III.9.3.3. CV measurements

Next we discuss the CV investigations done on the CuPc (samples *A* and *D*), CuPc/PCBM (*B*) and CuPc/POSS (*F*) incorporated APTMS films. On all samples the CV measurements were performed at four to five contacts with different diameters distributed over the surface of the sample, while the CV dependencies on one contact of each sample are presented in Fig. III.9.7a and 7b. The data are normalized to the contact area (A_C), the measurements shown in Fig. III.9.7 were performed on the smallest contact area of each sample. The thickness (*d*) of the samples were determined on different positions on the surface of the sample by AFM height histogram analysis as we described recently [202] (chapter III.7), where 167 nm, 190 nm, 20 nm, and 8 nm were evaluated as the thickness of the samples *A*, *B*, *D*, and *F*, respectively (see table III.9.5). In Fig. III.9.7a the CV data of thicker samples are compared. Although the sample *B* consists of a thicker dielectric it exhibits a higher value of the accumulation capacitance (C_{acc}). Therefore, as according to the plate capacitor approximation the capacitance depends inversely on the thickness, we can conclude that the dielectric constant (*k*) of sample *B* should be higher compared to sample *A*. In Fig. III.9.7b the CV data of the thinner layers are depicted. Here, the thinner layer of the sample *F* exhibits a higher C_{acc} value in comparison to the sample *D* (see Fig. III.9.7b). However, the thickness of the sample *F* is smaller by a factor of 2.5 than the thickness of sample *D*. Hence, as the C_{acc} value is not higher by this factor, a smaller *k* value of the layer *F* compared to the film *D* can be concluded qualitatively. In order to determine the permittivity of the films quantitatively we plotted the normalized C_{acc} values in dependence of the area according to Eq. III.7.3. Fitting the data in respect to Eq. III.7.3 the permittivity can easily be determined by the slope of the linear approximation. This is illustrated in Fig. III.9.8 where the normalized accumulation capacitance values of the samples *A*, *B*, *D*, and *F* as well as of a pure APTMS reference sample (chapter III.8.) are depicted. The results regarding the *k* values of the samples investigated in this work are overviewed in Table III.9.5. The error connected to the *k* value determination via the linear approximation was in the range of 1 % to 8 %. A probable additional interfacial SiO_2 capacitive contribution has not been taken into account. The samples *A* and *B* are relatively thick, therefore an influence on the permittivity is below 1 %, when a SiO_2 interfacial layer of 1 to 2 nm, as a typical value of the native oxide, is assumed. However it should be pointed out, that in the samples *D* and *F* the deviation of the permittivity is then in the range between 4 % and 13 %. Nevertheless, we observe a distinct decrease of the dielectric constant compared to the pure APTMS layer in all samples. While the dielectric constant of our pure APTMS film is 4.1 (chapter III.8), it reaches values of 2.6, 3.4, 3.3, and 1.8 for the samples *A*, *B*, *D*, and *F*, respectively. Discussing first the thicker layers we state that in particular the incorporation of CuPc of sufficient concentration into the APTMS film is effective for the permittivity decrease. In sample *A* only CuPc is incorporated into the APTMS and the CuPc content is higher (1.3 %) compared to sample *B* (0.7 %) where also PCBM (0.5 %) was added. Moreover, we realized a dielectric constant in the same range like for sample *B* when 1 % PCBM without CuPc was incorporated into APTMS (chapter III.7). But, if we include also the thinner layers into the discussion, we realize that an incorporation of only CuPc with a concentration lower than 0.1 % into the APTMS (sample *D*) matrix does not lead to such a strong decrease of the permittivity. However, the further addition of POSS molecules in a relatively low concentration (0.12 %) into the matrix while keeping the CuPc concentration in a similar range (sample *F*) delivers an outstanding value of the dielectric constant of the resulting APTMS based film. We argue that the additional incorporation of POSS molecules into the matrix induces the required porosity into the film. In our previous experiments (described in chapters III.7, III.8) we observed a similar behavior in the PCBM/POSS/APTMS system where also the combination of PCBM and POSS led to the best results in terms of the dielectric constant. However, there the amount of POSS monomers within the produced layers was overloaded leading to fractal patterns within the films surface (chapter III.8).

We have to note that we have not yet investigated heat treated films by CV. Products formed due to the hydrolysis and condensation reactions of the APTMS like –OH residual groups or water may still be present at the surface or within the film [134,202,210,215]. These species have strong permanent dipoles which may contribute to a higher overall polarizability of the final material and hence give rise to an increased dielectric constant [118,134,214]. Nevertheless, comparing the results of this work to our previous investigations (described in chapters III.7, III.8) we conclude that a CuPc/POSS material combination in a range as used in sample *F* in this work is most successful in order to reduce the final dielectric constant of APTMS based films incorporated by PCBM and/or POSS and/or CuPc molecules and might be a potential candidate for an ultra-low-*k* material.

Interestingly we observe in the thick APTMS layers mixed with CuPc (samples *A* and *B*, Fig. III.9.7a) much lower hysteresis inside the CV measurements in comparison to our former investigations, where C_{60}, PCBM or POSS species were incorporated into siloxane based films (chapters III.6, III.7, III.8). In chapter III.8 we have discussed a probable mechanism related to a filling and defilling of NH_3^+ charge trapping centers within the APTMS based films, where the trapping centers are caused by the interaction with ambient CO_2 (see also Equ. III.5.2), which is in agreement with other findings in literature [134]. We have further stated that these centers are mainly due to the APTMS itself (chapter III.8). If we compare the N 1s core level spectra in Fig. III.9.2 with that one of our previous work we find a clear reduction of the NH_3^+ related contribution (peak I) within the signals of the samples *A* and *B*. While the contribution of the NH_3^+ species to the

N 1s signal was over 11 % in our previous experiment (chapter III.8), in the samples *A* and *B* it equals around 4.1 % and 7.4 % respectively. Although the CuPc molecule reveals an insignificant Lewis acidity [216], the XPS spectra show, that CuPc in an adequate concentration (above or equal 0.7 %) within the APTMS seem to interact with the NH_3^+ centers, what results in a relatively low signal contribution of NH_3^+ species to the N 1s core level spectra. Contrary, the films with very low concentrations of CuPc (below 0.1 %, samples *D*, *E*, *F*) exhibit much stronger NH_3^+ signals (see Fig. III.9.6) which we observed also for the pure APTMS reference sample (chapter III.8). Moreover, the small remaining hysteresis in the CV data of the samples *A* and *B* is observed in the opposite direction compared to our previous investigations in chapters III.6, III.7, III.8. Therefore, we conclude that in the layers *A* and *B* NH_3^+ trapping centers are not the reason for the observed hysteresis. According to the direction of the hysteresis we propose here a charge injection coming from the Si substrate [134]. Probably both mechanisms are coexisting in all samples, but in the CuPc incorporated APTMS films the charge injection from the substrate is more prominent in contrast to the pure APTMS films as well as to the APTMS films mixed with C_{60}, PCBM and/or POSS, where the charge injection into the NH_3^+ centers is dominating. This fact is further supported by the hysteresis found in CV data of the thin layers *D* and *F* (Fig. III.9.7b). As discussed above the NH_3^+ contribution within the N 1s signal is much higher for these samples compared to the samples *A* and *B*. In agreement with this finding, we observe the same kind of hysteresis in the CV data of sample *F* like in the pure APTMS films or in the APTMS films added with C_{60}, PCBM or/and POSS molecules [202,215], (chapters III.7, III.8). However, in the CV measurement of sample *D*, we found the hysteresis in the other direction. But here we observed additionally a time dependence of the hysteresis, where at the beginning it was appearing in the same direction like in sample *F* but then it changed to the opposite direction like it is shown in the example curve in Fig. III. 9. 7b. This might be caused by a negative charging of the whole sample as we found a positive shift of the CV curve with time (inset in Fig. III.9.7b). This effect was not observed in all other samples. Due to the fact that annealing steps have not yet been performed moveable charges (Si–OH, water) might contribute to the hysteresis too [210,215].

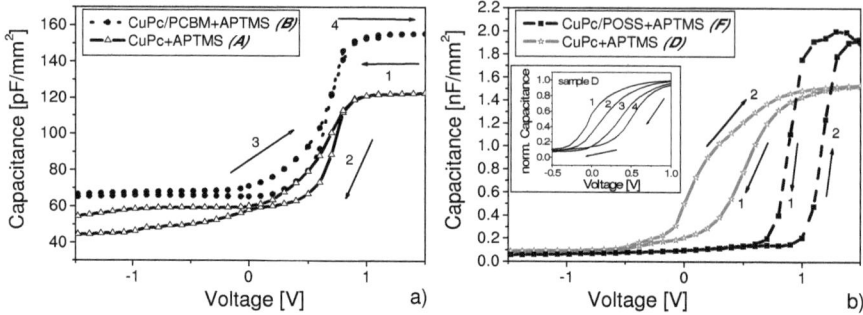

Fig. III.9.7.CV characteristics of samples *A* (part a, open triangles, solid line), *B* (part a, filled circles, dotted line), *D* (part b, open stars, gray line), and *F* (part b, filled rectangles, dashed line). The capacitance values were normalized to the contact areas. The measurement direction is indicated by indexed arrows. The inset in part b shows repeated measurements from accumulation to inversion for sample *D*.

Additionally we observe a positive shift of the flatband voltage in respect to the voltage axis for the samples *A, B,* and *F* (sample *D* is not further discussed here because of the above mentioned charge injection with time). According to our previous report in chapter III. 8 the ideal flatband voltage originating from the work function difference between the silver metal and the semiconductor should be in the range of -0.2 V to 0 V for these samples [215]. The measured flatband voltage of these samples is in the range between 0.74 V and 0.81 V for the samples *A* and *B* and between 0.96 V and 1.01 V for sample *F* corresponding to negative fixed oxide charges in the range of 5.7-$7.7*10^{10}$ cm^{-2}, 7.2-$9.7*10^{10}$ cm^{-2}, and 1.2-$1.5*10^{12}$ cm^{-2} for the samples *A, B,* and *F,* respectively [38].

Recently, we discussed carbamate ($NHCOO^-$) species in our APTMS based films originating from interactions with ambient CO_2 (chapters III.7, III.8) which might be responsible for the flatband voltage shift in positive direction (chapter III.8). Also here in the C 1s core level spectra of the samples *A, B* and *F* we observe $NHCOO^-$ species (peak I in the related signals in Figs. III.9.2 and 6). Because the film *F* is not exhibiting a higher $NHCOO^-$ contribution inside the C 1s data compared to the samples *A* and *B* we propose that an additional charge injection leading to negative fixed charges took place in the beginning of the measurement due to the low thickness of the film.

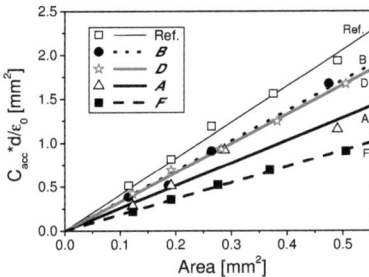

Fig. III.9.8. Normalized accumulation capacitance versus the contact area of the samples A (open triangle), B (filled circles), D (open stars), F (filled rectangles), and a reference sample of pure APTMS taken from our previous investigation (open squares, chapter III. 8). The data are normalized corresponding to Eq. III.7.3. Linear fits to the data are performed according to Eq. III.7.3 (A: thick solid line, B: dotted line, D: gray line, F: dashed line, reference sample: thin solid line). The results of the fits are discussed in the text and displayed in Table III.9.5.

Table III.9.5. Initial compositions and thicknesses of the electrically investigated samples and relative dielectric constant deduced by fits according to Equ. III.7.3.

Sample	APTMS [%]	CuPc [%]	POSS [%]	PCBM [%]	Thickness [nm]	k
Ref. [215]	100	0	0	0	138	4.1
A	98.7	1.3	0	0	167	2.6
B	98.8	0.7	0	0.5	190	3.4
D	99.91	0.09	0	0	20	3.3
F	99.84	0.04	0.12	0	8	1.8

III.9.4. Conclusions

Basing on the collected spectroscopic data one may observe, that even low concentration of the CuPc and POSS dopants (in the range of 0.04 %-1.3 %) influences the Si: N ratio within the APTMS layers sufficiently to be detected by XPS quantitative analysis. XPS revealed that the

POSS containing layers were the least affected by ambient influence while at the same time the CuPc molecules accelerate the surface oxidation process.

The CV measurements indicated a distinct decrease of the dielectric constant of all composite APTMS based films described in this chapter compared to the pure APTMS layer. The incorporation of POSS and CuPc dopants in a proper concentration (in the range of 0.1 %-0.2 %) into the aminopropylsiloxane matrix allowed to obtain the dielectric constant value of 1.8. Although up to now CuPc oligomers have been applied mainly as a high dielectric constant fillers in a polymer matrix [58,59], our investigation demonstrates, that an appropriate chosen CuPc/POSS dopant combination seems to deliver a successful reduction of the final dielectric constant of APTMS based films. Moreover, CuPc molecules seem to decrease the hysteresis inside the CV measurements of the investigated samples probably due to an interaction with NH_3^+ trapping centers.

IV. Conclusions and outlook

A complex investigation concerning the processing and the possible application of thin, carbon species reach films has been performed. As the experiments revealed, the combination of different, complementary investigation techniques is an indisputable advantage allowing the gathering of a wide spectrum of information about material properties. This leads consequently to the continuous improvements of the desired properties of the produced material.

The combined spectroscopic methods (XPS, NEXAFS and FTIR) revealed the possible changes within the fullerene structure upon processing in the laboratory conditions. The stability of many sorts of the fullerene samples deposited by means of several techniques has been intensively investigated. The data reported in this thesis regarding toluene treated fullerene revealed the significant increase of the sp^3 hybridized carbon species in comparison to the C_{60} deposited by sublimation. In addition these data exhibited the presence of carbon atoms attached to different oxygen-containing moieties, with the superiority of the C–O species. The results of the experiments clearly indicated that the layers prepared and measured *in situ* under vacuum conditions exhibit the best properties. This observation confirmed the theory that *in situ* methods like ALD provide the opportunity to obtain layers of best quality. On the other hand these techniques also have limitations, for instance they require a very precise temperature control. This factor, as it has been confirmed in this thesis, is crucial in the aminopropylsiloxane films processing while even relatively small temperature deviation (in the range 50 °C) may decide whether the deposition process succeed. Therefore the majority of the investigated samples have been prepared by spin-coating.

As a curiosity, the stability of the low concentrated fullerol water solution in ambient conditions has been investigated. XPS and NEXAFs revealed, that after one year of the storage in the air without protection from the light the 0.01% solution still –C=C– species which were not affected by the oxidation process. This phenomenon has been explained by the occurrence of the kinetic limitation due to either a steric hindrance or the agglomerates formation, what would promote the passivation process.

A considerable attention has been paid to the films obtained through the interactions of APTMS functional groups with carbon species. The combination of these components opened an opportunity to obtain a rich spectrum of hybrid composite materials possessing interesting chemical, morphological and electrical properties. First, the properties of composites obtained by combined spin-coating and evaporation techniques have been investigated by means of spectroscopic and microscopic methods. Spectroscopy measurements confirmed the high affinity of the fullerene molecules to the amino groups while AFM revealed that fullerene diameter is not affected by the interactions with APTMS upon performed procedure. However also time-related significant changes within such layer structures related to the amine group migrations have been observed.

The APTMS based composite films obtained by spin-coating revealed a significantly higher stability against the ambient influence. Moreover, the films prepared while dispersing the C_{60} within the APTMS monomer exhibit higher homogeneity than the fullerene containing layer prepared by sol-gel method and had a relatively low dielectric constant (below 4.0). This observation gave the opportunity of further improvements of this kind of composite material by varying the dopant species and concentration.

From this point, successive steps leading to the decreasing of the resulting permittivity of the hybrid material and at the same time improving the films quality have been shown. The replacement of C_{60} by its derivative PCBM allowed the increase of the overall carbon species concentration within the resulting material. The introduction of POSS molecules as an additional dopant improved the surface resistance against the ambient influence and at the same time decreased the permittivity of the deposited layer. When the POSS concentration was too high, the films were too inhomogeneous to perform the CV investigation. However, the resulting agglomeration process, explained either by a physical aggregation via a phase separation or by a chemical POSS–POSS condensation reaction, led to the formation of an original fractal-shaped cluster.

During the further researches aimed in the direction of the ultra-low k material preparation, CuPc has been applied as an additional dopant of the produced material. After a series of investigation of the APTMS based hybrid materials containing single and combined dopants as well, the best quality revealed a film containing the properly chosen low concentration of CuPc and POSS molecules dispersed within the siloxane matrix. The layer obtained in this way revealed an ultra-low dielectric constant of 1.8, high homogeneity and a lower hysteresis inside the CV measurements compared to the previously investigated films what has been explained by an interaction between CuPc and NH_3^+ trapping centers formed as a result of ambient influence on APTMS based network. However, a series of additional investigations, including mechanical testing, would be necessary in order to verify the possibility of the obtained material low k application.

The performed experiments open a wide range of possible improvements and developments. First of all the homogeneity of the low-k layers could be further improved while introducing the properly chosen solvent and surfactant to the spin-coated substance. An external catalyst could enable a better control of the reactions taking place within the material. Regarding this topic, it would be very useful to perform also wider CV investigation of the dependence between surface roughness and the dielectric constant of the material.

As revealed by the investigations performed in chapter III.4 the heat treatment of the composite materials allows to obtain a silicon network enriched with carbon species coming from the applied dopants. The properly chosen heat treatment factors (including time and temperature) could lead to a carbon reach material of the desired porosity created by the gaseous by-products of the process. The application of the surfactants and some external catalyst could allow the control of such material homogeneity.

Finally, the experiment could be further developed with the support of the ALD technique. The main problems to overcome in this situation would be a proper temperature control. Also the solution leading to a sufficient deposition rate of the film ought to be found.

References

[1] B. Hu, S.-H. Yu, K. Wang, L. Liu, X.-W. Xu, Functional carbonaceous materials from hydrothermal carbonization of biomass: an effective chemical process, Dalton Trans. (2008) 5414.
[2] L. Dai, Carbon nanotechnology: recent developments in chemistry, physics, materials science and device applications, first ed., Elsevier Science, 2006.
[3] P.M. Ajayan, J.M. Tour, Nanotube composites, Nature. 447 (2007) 1066.
[4] K.-C. Chang, C.-Y. Lin, H.-F. Lin, S.-C. Chiou, W.-C. Huang, J.-M. Yeh, et al., Thermally and mechanically enhanced epoxy resin-silica hybrid materials containing primary amine-modified silica nanoparticles, J. Appl. Polym. Sci. 108 (2008) 1629.
[5] N.Z. Muradov, T.N. Veziroglu, From hydrocarbon to hydrogen-carbon to hydrogen economy, International Journal of Hydrogen Energy. 30 (2005) 225.
[6] J.C.F. Rodríguez-Reyes, A.V. Teplyakov, Chemistry of Organometallic Compounds on Silicon: The First Step in Film Growth, Chem. Eur. J. 13 (2007) 9164.
[7] G.E. Moore, No exponential is forever: but "Forever" can be delayed!, Solid-State Circuits Conference, 2003. Digest of Technical Papers. ISSCC. 2003 IEEE International, 2003: pp. 20 vol.1.
[8] P. de Rouffignac, Z. Li, R.G. Gordon, Sealing Porous Low-k Dielectrics with Silica, Electrochem. Solid-State Lett. 7 (2004) G306.
[9] K. Maex, M.R. Baklanov, D. Shamiryan, F. Iacopi, S.H. Brongersma, Z.S. Yanovitskaya, Low dielectric constant materials for microelectronics, J. Appl. Phys. 93 (2003) 8793.
[10] M.R. Baklanov, K.P. Mogilnikov, Non-destructive characterisation of porous low-k dielectric films, Microelectron. Eng. 64 (2002) 335.
[11] S.-K. Kwak, K.-H. Jeong, S.-W. Rhee, Nanocomposite Low-k SiCOH Films by Direct PECVD Using Vinyltrimethylsilane, J. Electrochem. Soc. 151 (2004) F11.
[12] R.R. Sahoo, A. Patnaik, Binding of fullerene C_{60} to gold surface functionalized by self-assembled monolayers of 8-amino-1-octane thiol: a structure elucidation, J. Colloid Interface Sci. 268 (2003) 43.
[13] A. Hirsch, M. Brettreich, Fullerenes: Chemistry and Reactions, first ed., Wiley-VCH, Weinheim, 2005.
[14] D.Q. Li, B.I. Swanson, Surface acoustic wave thin-film chemical microsensors based on covalently bound C_{60} derivatives: a molecular self-assembly approach, Langmuir. 9 (1993) 3341.
[15] K.M. Kadish, E.S.F. Group, Fullerenes, first ed., The Electrochemical Society, 1999.
[16] H. Hertz, Annalen Der Physik. 31 (1887) 982.
[17] A. Einstein, Annalen Der Physik. 17 (1905) 132.
[18] Y. Gao, Surface analytical studies of interfaces in organic semiconductor devices, Materials Science and Engineering: R: Reports. 68 (2010) 39.
[19] K. Siegbahn, C. Nordling, A. Fahlman, R. Nordberg, K. Hamrin, J. Hedman, et al., ESCA, first ed., Almqvist & Wiksells Boktryckeri AB, 1967.
[20] M.P. Seah, I.S. Gilmore, S.J. Spencer, Quantitative XPS: I. Analysis of X-ray photoelectron intensities from elemental data in a digital photoelectron database, J. Electron Spectrosc. Relat. Phenom. 120 (2001) 93.
[21] M.P. Seah, I.S. Gilmore, S.J. Spencer, XPS: Binding Energy Calibration of Electron Spectrometers 4-Assessment of Effects for Different X-ray Sources, Analyser Resolutions, Angles of Emission and Overall Uncertainties, Surf. Interface Anal. 26 (1998) 617.
[22] M. Repoux, Comparison of background removal methods for XPS, Surf. Interface Anal. 18 (1992) 567.
[23] M.V. Baker, J.D. Watling, Functionalization of Alkylsiloxane Monolayers via Free-Radical Bromination, Langmuir. 13 (1997) 2027.
[24] J. Du, P. Zhang, Composition and Microstructure of Magnetron Sputtering Deposited Ti-containing Amorphous Carbon Films, J. Mater. Sci. Technol. 23 (2007) 571.
[25] J.F. Moulder, W.F. Stickle, P.E. Sobol, Handbook of X Ray Photoelectron Spectroscopy, Perkin-Elmer, Physical Electronics Division, Eden Prairie, 1993.

[26] A. Kuznetsova, I. Popova, Yates, M.J. Bronikowski, C.B. Huffman, J. Liu, et al., Oxygen-Containing Functional Groups on Single-Wall Carbon Nanotubes: NEXAFS and Vibrational Spectroscopic Studies, Journal of the American Chemical Society. 123 (2001) 10699.
[27] D. Schmeisser, M. Tallarida, K. Henkel, K. Müller, D. Mandal, D. Chumakov, et al., Characterization of oxidic and organic materials with synchrotron radiation based XPS and XAS, Mater.Sci.-Poland. 27 (2009) 141.
[28] J. Stohr, Nexafs Spectroscopy, Springer Series in Surface Science, ed., Springer-Verlag, New York, 1996.
[29] W. Gudat, C. Kunz, Close Similarity between Photoelectric Yield and Photoabsorption Spectra in the Soft-X-Ray Range, Phys. Rev. Lett. 29 (1972) 169.
[30] A.J. Achkar, T.Z. Regier, H. Wadati, Y.-J. Kim, H. Zhang, D.G. Hawthorn, Bulk sensitive x-ray absorption spectroscopy free of self-absorption effects, Phys. Rev. B. 83 (2011) 081106.
[31] J. Jaklevic, J.A. Kirby, M.P. Klein, A.S. Robertson, G.S. Brown, P. Eisenberger, Fluorescence detection of exafs: Sensitivity enhancement for dilute species and thin films, Solid State Communications. 23 (1977) 679.
[32] D. Schmeisser, P. Hoffmann, G. Beuckert, Electronic Properties of the Interface Formed by Pr 2 O 3 Growth on Si (001), Si (111) and SiC (0001) Surfaces, first ed., Springer-Verlag, London, 2005.
[33] P.R. Griffiths, J.A.D. Haseth, Fourier transform infrared spectrometry, Wiley-Interscience, 2007.
[34] D.L. Pavia, G.M. Lampman, Introduction to spectroscopy, Cengage Learning, 2009.
[35] F.J. Giessibl, AFM's path to atomic resolution, Materials Today. 8 (2005) 32.
[36] R. Garcia, R. Perez, Dynamic atomic force microscopy methods, Surface Science Reports. 47 (2002) 197.
[37] D.Y. Abramovitch, S.B. Andersson, L.Y. Pao, G. Schitter, A tutorial on the mechanisms, dynamics, and control of atomic force microscopes, in: American Control Conference, 2007. ACC'07, 2007: pp. 3488.
[38] S.M. Sze, Physics of Semiconductor Devices, second ed., John Wiley & Sons, New York, 1981.
[39] K. Henkel, Electrical Investigations on Praseodymium Oxide/Aluminum Oxynitride Containing Metal–Insulator–Semiconductor Stacks and on Metal–Ferroelectric–Insulator–Semiconductor Structures Consisting of Poly(vinylidene fluoride trifluoroethylene), Ph.D. Thesis, Brandenburg University of Technology, Germany, Shaker-Verlag, 2009.
[40] B.S. Sherigara, W. Kutner, F. D'Souza, Electrocatalytic Properties and Sensor Applications of Fullerenes and Carbon Nanotubes, Electroanalysis. 15 (2003) 753.
[41] G. Yu, J. Gao, J.C. Hummelen, F. Wudl, A.J. Heeger, Polymer Photovoltaic Cells: Enhanced Efficiencies via a Network of Internal Donor-Acceptor Heterojunctions, Science. 270 (1995) 1789 .
[42] R.C. Haddon, A.S. Perel, R.C. Morris, T.T.M. Palstra, A.F. Hebard, R.M. Fleming, C_{60} thin film transistors, Appl. Phys. Lett. 67 (1995) 121.
[43] G. Lu, L. Li, X. Yang, Creating a Uniform Distribution of Fullerene C_{60} Nanorods in a Polymer Matrix and its Photovoltaic Applications, Small. 4 (2008) 601.
[44] H.W. Kroto, A.W. Allaf, S.P. Balm, C_{60}: Buckminsterfullerene, Chemical Reviews. 91 (1991) 1213.
[45] F. Wudl, The chemical properties of buckminsterfullerene (C_{60}) and the birth and infancy of fulleroids, Accounts of Chemical Research. 25 (1992) 157.
[46] H. Okamura, T. Takemura, M. Tsunooka, M. Shirai, Synthesis of Novel C_{60}-containing Polymers Based on Poly(vinyl phenol) and Their Photo-transformation Properties, Polym. Bull. 52 (2004) 381.
[47] L.J. Terminello, D.K. Shuh, F.J. Himpsel, D.A. Lapiano-Smith, J. Stöhr, D.S. Bethune, et al., Unfilled orbitals of C_{60} and C_{70} from carbon K-shell X-ray absorption fine structure, Chemical

Physics Letters. 182 (1991) 491.
[48] R.C. Haddon, L.E. Brus, K. Raghavachari, Electronic structure and bonding in icosahedral C_{60}, Chemical Physics Letters. 125 (1986) 459.
[49] B. Ma, A.M. Milton, Y.-P. Sun, Infrared spectroscopy of all-carbon poly[60]fullerene polymer and [60]fullerene dimer model, Chemical Physics Letters. 288 (1998) 854.
[50] M.S. Dresselhaus, G. Dresselhaus, P.C. Eklund, Raman Scattering in Fullerenes, Journal of Raman Spectroscopy. 27 (1996) 351.
[51] D. Sun, C.A. Reed, Crystal engineering a linear polymer of C_{60} fullerene via supramolecular pre-organization, Chem. Commun. (2000) 2391.
[52] D. Bonifazi, O. Enger, F. Diederich, Supramolecular [60]fullerene chemistry on surfaces, Chem. Soc. Rev. 36 (2007) 390.
[53] F. Langa, J.-F. Nierengarten, Fullerenes: principles and applications, first ed., Royal Society of Chemistry, Cambridge, 2007.
[54] A. Gulino, S. Bazzano, G.G. Condorelli, S. Giuffrida, P. Mineo, C. Satriano, et al., Engineered Silica Surfaces with an Assembled C_{60} Fullerene Monolayer, Chem. Mater. 17 (2005) 1079.
[55] K.T. Park, A. Miller, K. Klier, R.L. Opila, J.E. Rowe, Heteroepitaxial copper phthalocyanine on Au(001) studied by high-resolution X-ray photoelectron spectroscopy, Surface Science. 529 (2003) L285.
[56] H. Peisert, T. Schwieger, J.M. Auerhammer, M. Knupfer, M.S. Golden, J. Fink, et al., Order on disorder: Copper phthalocyanine thin films on technical substrates, J. Appl. Phys. 90 (2001) 466.
[57] L.-C. Hu, M. Khiterer, S.-J. Huang, J.C.C. Chan, J.R. Davey, K.J. Shea, Uniform, Spherical Bridged Polysilsesquioxane Nano- and Microparticles by a Nonemulsion Method, Chem. Mater. 22 (2010) 5244.
[58] Q.M. Zhang, H. Li, M. Poh, F. Xia, Z.-Y. Cheng, H. Xu, et al., An all-organic composite actuator material with a high dielectric constant, Nature. 419 (2002) 284.
[59] N. Shi, R. Ramprasad, Dielectric properties of Cu-phthalocyanine systems from first principles, Appl. Phys. Lett. 89 (2006) 102904.
[60] X. Zhang, Y. Huang, Y. Wang, Y. Ma, Z. Liu, Y. Chen, Synthesis and characterization of a graphene-C_{60} hybrid material, Carbon. 47 (2009) 334.
[61] K. Zagorodniy, H. Hermann, M. Taut, Structure and properties of computer-simulated fullerene-based ultralow- k dielectric materials, Phys. Rev. B. 75 (2007) 245430.
[62] J. Zhang, H. Wang, X. Yan, J. Wang, J. Shi, D. Yan, Phthalocyanine composites as high-mobility semiconductors for organic thin-film transistors, Advanced Materials. 17 (2005) 1191.
[63] X. Zhang, A.V. Teplyakov, Adsorption of C_{60} Buckminster Fullerenes on an 11-Amino-1-undecene-Covered Si(111) Substrate, Langmuir. 24 (2008) 810.
[64] Y.-G. Guo, L.-J. Wan, C.-J. Li, D.-M. Chen, C. Wang, C.-R. Wang, et al., The effects of annealing on the structures and electrical conductivities of fullerene-derived nanowires, J. Mater. Chem. 14 (2004) 914.
[65] B. Adolphi, O. Berger, W.J. Fischer, Angle-resolved XPS measurements on copper phthalocyanine thin films, Applied Surface Science. 179 (2001) 103.
[66] Y. Niwa, X-ray photoelectron spectroscopy of tetraphenylporphin and phthalocyanine, J. Chem. Phys. 60 (1974) 799.
[67] V. Datsyuk, M. Kalyva, K. Papagelis, J. Parthenios, D. Tasis, A. Siokou, et al., Chemical oxidation of multiwalled carbon nanotubes, Carbon. 46 (2008) 833.
[68] E.T. Kang, K.G. Neoh, K.L. Tan, B.T.G. Tan, XPS studies of charge transfer interactions in some polyphenylacetylene-electron acceptor systems, Journal of Polymer Science Part B: Polymer Physics. 27 (1989) 2061.
[69] G. Dufour, C. Poncey, F. Rochet, H. Roulet, M. Sacchi, M. De Santis, et al., Copper phthalocyanine on Si(111)-7×7 and Si(001)-2×1 surfaces: an X-ray photoemission spectroscopy and synchrotron X-ray absorption spectroscopy study, Surface Science. 319 (1994) 251.
[70] C. Jama, A. Al khawwam, A.-S. Loir, P. Goudmand, O. Dessaux, L. Gengembre, et al., X-

ray photoelectron spectroscopy study of carbon nitride coatings deposited by IR laser ablation in a remote nitrogen plasma atmosphere, Surf. Interface Anal. 31 (2001) 815.
[71] T.I.T. Okpalugo, P. Papakonstantinou, H. Murphy, J. McLaughlin, N.M.D. Brown, High resolution XPS characterization of chemical functionalised MWCNTs and SWCNTs, Carbon. 43 (2005) 153.
[72] K. Yuan, Z.F. Li, L.L. LÜ, X.N. Shi, Synthesis and characterization of well-defined polymer brushes grafted from silicon surface via surface reversible addition–fragmentation chain transfer (RAFT) polymerization, Materials Letters. 61 (2007) 2033.
[73] R. Machorro, E.C. Samano, G. Soto, L. Cota, SiCxNy thin films alloys prepared by pulsed excimer laser deposition, Appl. Surf. Sci. 127-129 (1998) 564.
[74] A. Ruocco, F. Evangelista, R. Gotter, A. Attili, G. Stefani, Evidence of Charge Transfer at the Cu-phthalocyanine/Al(100) Interface†, The Journal of Physical Chemistry C. 112 (2008) 2016.
[75] H. Ago, T. Kugler, F. Cacialli, W.R. Salaneck, M.S.P. Shaffer, A.H. Windle, et al., Work Functions and Surface Functional Groups of Multiwall Carbon Nanotubes, The Journal of Physical Chemistry B. 103 (1999) 8116.
[76] J.T. Titantah, D. Lamoen, sp^3/sp^2 characterization of carbon materials from first-principles calculations: X-ray photoelectron versus high energy electron energy-loss spectroscopy techniques, Carbon. 43 (2005) 1311.
[77] R. Haerle, E. Riedo, A. Pasquarello, A. Baldereschi, sp^2/sp^3 hybridization ratio in amorphous carbon from C 1s core-level shifts: X-ray photoelectron spectroscopy and first-principles calculation, Phys. Rev. B. 65 (2001) 045101.
[78] S. Sawamura, N. Fujita, High-pressure solubility of fullerene C_{60} in toluene, Carbon. 45 (2007) 965.
[79] H. Hoffmann, F. Zaera, R. Mark Ormerod, R.M. Lambert, Lu Ping Wang, W.T. Tysoe, Discovery of a tilted form of benzene chemisorbed on Pd(111): As NEXAFS and photoemission investigation, Surface Science. 232 (1990) 259.
[80] R. Taylor, J.P. Parsons, A.G. Avent, S.P. Rannard, T.J. Dennis, J.P. Hare, et al., Degradation of C_{60} by light, Nature. 351 (1991) 277.
[81] J.A. Leiro, M.H. Heinonen, T. Laiho, I.G. Batirev, Core-level XPS spectra of fullerene, highly oriented pyrolitic graphite, and glassy carbon, Journal of Electron Spectroscopy and Related Phenomena. 128 (2003) 205.
[82] T. Ohwaki, H. Ishida, Comparison Between FTIR - and XPS - Characterization of Carbon Fiber Surfaces, The Journal of Adhesion. 52 (1995) 167.
[83] T. Ressler, WinXAS: a Program for X-ray Absorption Spectroscopy Data Analysis under MS-Windows, J Synchrotron Rad. 5 (1998) 118.
[84] D.A. Outka, J. Stöhr, Curve fitting analysis of near-edge core excitation spectra of free, adsorbed, and polymeric molecules, J. Chem. Phys. 88 (1988) 3539.
[85] S. Bernard, O. Beyssac, K. Benzerara, N. Findling, G. Tzvetkov, G.E. Brown Jr., XANES, Raman and XRD study of anthracene-based cokes and saccharose-based chars submitted to high-temperature pyrolysis, Carbon. 48 (2010) 2506.
[86] H.-K. Jeong, L. Colakerol, M.H. Jin, P.-A. Glans, K.E. Smith, Y.H. Lee, Unoccupied electronic states in graphite oxides, Chemical Physics Letters. 460 (2008) 499.
[87] A. Braun, A. Kubatova, S. Wirick, S.B. Mun, Radiation damage from EELS and NEXAFS in diesel soot and diesel soot extracts, Journal of Electron Spectroscopy and Related Phenomena. 170 (2009) 42.
[88] T. Okajima, K. Teramoto, R. Mitsumoto, H. Oji, Y. Yamamoto, I. Mori, et al., Polarized NEXAFS Spectroscopic Studies of Poly(butylene terephthalate), Poly(ethylene terephthalate), and Their Model Compounds, J. Phys. Chem. A. 102 (1998) 7093.
[89] S. Saxena, T.A. Tyson, E. Negusse, Investigation of the Local Structure of Graphene Oxide, The Journal of Physical Chemistry Letters. 1 (2010) 3433.
[90] O. Dhez, H. Ade, S.G. Urquhart, Calibrated NEXAFS spectra of some common polymers,

J. Electron Spectrosc. Relat. Phenom. 128 (2003) 85.
[91] C. Lenardi, P. Piseri, V. Briois, C.E. Bottani, A.L. Bassi, P. Milani, Near-edge x-ray absorption fine structure and Raman characterization of amorphous and nanostructured carbon films, J. Appl. Phys. 85 (1999) 7159.
[92] H. Oji, R. Mitsumoto, E. Ito, H. Ishii, Y. Ouchi, K. Seki, et al., Core hole effect in NEXAFS spectroscopy of polycyclic aromatic hydrocarbons: Benzene, chrysene, perylene, and coronene, J. Chem. Phys. 109 (1998) 10409.
[93] W. Chen, H. Huang, S. Chen, X.Y. Gao, A.T.S. Wee, Low-Temperature Scanning Tunneling Microscopy and Near-Edge X-ray Absorption Fine Structure Investigations of Molecular Orientation of Copper(II) Phthalocyanine Thin Films at Organic Heterojunction Interfaces, J. Phys. Chem. C. 112 (2008) 5036.
[94] M. Wohlers, A. Bauer, T. Rühle, F. Neitzel, H. Werner, R. Schlögl, The Dark Reaction of C_{60} and of C_{70} with Molecular Oxygen at Atmospheric Pressure and Temperatures between 300 K and 800 K, Fullerene Science and Technology. 5 (1997) 49.
[95] D.S. Bethune, G. Meijer, W.C. Tang, H.J. Rosen, W.G. Golden, H. Seki, et al., Vibrational Raman and infrared spectra of chromatographically separated C_{60} and C_{70} fullerene clusters, Chemical Physics Letters. 179 (1991) 181.
[96] G.V. Andrievsky, V.K. Klochkov, A.B. Bordyuh, G.I. Dovbeshko, Comparative analysis of two aqueous-colloidal solutions of C_{60} fullerene with help of FTIR reflectance and UV-Vis spectroscopy, Chemical Physics Letters. 364 (2002) 8.
[97] S.C. Mojumdar, Preparation, thermal, spectral and microscopic studies of calcium silicate hydrate - poly(acrylic acid) nanocomposite materials, Journal of Thermal Analysis and Calorimetry. 85 (2006) 99.
[98] V. Zucolotto, M. Ferreira, M.R. Cordeiro, C.J.L. Constantino, D.T. Balogh, A.R. Zanatta, et al., Unusual Interactions Binding Iron Tetrasulfonated Phthalocyanine and Poly(allylamine hydrochloride) in Layer-by-Layer Films, The Journal of Physical Chemistry B. 107 (2003) 3733.
[99] N. Primeau, C. Vautey, M. Langlet, The effect of thermal annealing on aerosol-gel deposited SiO_2 films: a FTIR deconvolution study, Thin Solid Films. 310 (1997) 47.
[100] C.Y. Liang, R.H. Marchessault, Infrared spectra of crystalline polysaccharides. II. Native celluloses in the region from 640 to 1700 cm^{-1}, Journal of Polymer Science. 39 (1959) 269.
[101] N. Sundaraganesan, H. Saleem, S. Mohan, Vibrational spectra, assignments and normal coordinate analysis of 3-aminobenzyl alcohol, Spectrochimica Acta Part A: Molecular and Biomolecular Spectroscopy. 59 (2003) 2511.
[102] G.H. Kroll, P.J. Benning, Y. Chen, T.R. Ohno, J.H. Weaver, L.P.F. Chibante, et al., Interaction of O2 with C_{60}: photon-induced oxidation, Chemical Physics Letters. 181 (1991) 112.
[103] Y. Liu, F. Jiao, Y. Qiu, W. Li, Y. Qu, C. Tian, et al., Immunostimulatory properties and enhanced TNF-α mediated cellular immunity for tumor therapy by $C_{60}(OH)_{20}$ nanoparticles, Nanotechnology. 20 (2009) 415102.
[104] B. Vileno, P.R. Marcoux, M. Lekka, A. Sienkiewicz, T. Fehér, L. Forró, Spectroscopic and Photophysical Properties of a Highly Derivatizated C_{60} Fullerol, Advanced Functional Materials. 16 (2006) 120.
[105] L.O. Husebo, B. Sitharaman, K. Furukawa, T. Kato, L.J. Wilson, Fullerenols Revisited as Stable Radical Anions, Journal of the American Chemical Society. 126 (2004) 12055.
[106] M.E. Rincón, R.A. Guirado-López, J.G. Rodríguez-Zavala, M.C. Arenas-Arrocena, Molecular films based on polythiophene and fullerol: theoretical and experimental studies, Solar Energy Materials and Solar Cells. 87 (2005) 33.
[107] T.P. Martin, U. Näher, H. Schaber, U. Zimmermann, Clusters of fullerene molecules, Phys. Rev. Lett. 70 (1993) 3079.
[108] K.J. Kitching, D.J. Wilson, P.J. Doherty, R.L. Williams, An X-ray Photoelectron Spectroscopy study of biomedical polyurethane modified using low-power plasma, Journal of X-Ray Science and Technology. 9 (1999) 77.
[109] J.A. Nisha, V. Sridharan, J. Janaki, Y. Hariharan, V.S. Sastry, C.S. Sundar, et al., Studies of

C_{60} Oxidation and Products, The Journal of Physical Chemistry. 100 (1996) 4503.
[110] D. Kondo, K. Sakamoto, H. Takeda, F. Matsui, T. Ohta, K. Amemiya, et al., Thermal effect in unoccupied molecular orbitals of C_{60} molecules adsorbed on a Si (001)-(2 x 1) surface studied by NEXAFS, J. Synchrotron Radiat. 8 (2001) 505.
[111] S. di Stasio, A. Braun, Comparative NEXAFS Study on Soot Obtained from an Ethylene/Air Flame, a Diesel Engine, and Graphite, Energy Fuels. 20 (2006) 187.
[112] R.R. Cooney, S.G. Urquhart, Chemical Trends in the Near-Edge X-ray Absorption Fine Structure of Monosubstituted and Para-Bisubstituted Benzenes, J. Phys. Chem. B. 108 (2004) 18185.
[113] R.M. Petoral, K. Uvdal, NEXAFS study of amino acid analogues assembled on gold, Physica Scripta. 115 (2005) 851.
[114] S. Ek, E.I. Iiskola, L. Niinistö, J. Vaittinen, T.T. Pakkanen, J. Keränen, et al., Atomic Layer Deposition of a High-Density Aminopropylsiloxane Network on Silica through Sequential Reactions of γ-Aminopropyltrialkoxysilanes and Water, Langmuir. 19 (2003) 10601.
[115] J. Bachmann, R. Zierold, Y.T. Chong, R. Hauert, C. Sturm, R. Schmidt-Grund, et al., A Practical, Self-Catalytic, Atomic Layer Deposition of Silicon Dioxide, Angew. Chem. Int. Ed. 47 (2008) 6177.
[116] H.K. Kim, J.P. Lee, C.R. Park, H.T. Kwak, M.M. Sung, Thermal Decomposition of Alkylsiloxane Self-Assembled Monolayers in Air, The Journal of Physical Chemistry B. 107 (2003) 4348.
[117] H.G. Linde, R.T. Gleason, Thermal stability of the silica-aminopropylsilane-polyimide interface, Journal of Polymer Science: Polymer Chemistry Edition. 22 (1984) 3043.
[118] Y. Li, I. Ciofi, L. Carbonell, N. Heylen, J. Van Aelst, M.R. Baklanov, et al., Influence of absorbed water components on SiOCH low-k reliability, J. Appl. Phys. 104 (2008) 034113.
[119] H.L. Cabibil, V. Pham, J. Lozano, H. Celio, R.M. Winter, J.M. White, Self-Organized Fibrous Nanostructures on Poly[(aminopropyl)siloxane] Films Studied by Atomic Force Microscopy, Langmuir. 16 (2000) 10471.
[120] M.P. Felicissimo, D. Jarzab, M. Gorgoi, M. Forster, U. Scherf, M.C. Scharber, et al., Determination of vertical phase separation in a polyfluorene copolymer: fullerene derivative solar cell blend by X-ray photoelectron spectroscopy, J. Mater. Chem. 19 (2009) 4899.
[121] A.P. Dementjev, A. De Graaf, M.C.M. Van de Sanden, K.I. Maslakov, A.V. Naumkin, A.A. Serov, X-Ray photoelectron spectroscopy reference data for identification of the C3N4 phase in carbon–nitrogen films, Diamond Relat. Mater. 9 (2000) 1904.
[122] V. Datsyuk, C. Guerret-Piécourt, S. Dagréou, L. Billon, J.-C. Dupin, E. Flahaut, et al., Double walled carbon nanotube/polymer composites via in-situ nitroxide mediated polymerisation of amphiphilic block copolymers, Carbon. 43 (2005) 873.
[123] I. George, P. Viel, C. Bureau, J. Suski, G. Lécayon, Study of the Silicon/γ-APS/Pyralin Assembly Interfaces by X-ray Photoelectron Spectroscopy, Surf. Interface Anal. 24 (1996) 774.
[124] T. Thärigen, D. Mayer, R. Hesse, P. Streubel, D. Lorenz, P. Grau, et al., XANES and XPS characterization of hard amorphous CSi_xN_y thin films grown by RF nitrogen plasma assisted pulsed laser deposition, Fresenius J. Anal. Chem. 365 (1999) 244.
[125] X. Zhang, H. Yang, F. Zhang, K.-Y. Chan, Preparation and characterization of Pt-TiO_2-SiO_2 mesoporous materials and visible-light photocatalytic performance, Materials Letters. 61 (2007) 2231.
[126] H. He, N. Swami, B.E. Koel, Reaction of C_{60} with oxygen adatoms on Pt(111...), J. Chem. Phys. 110 (1999) 1173.
[127] M. Ramm, M. Ata, K.W. Brzezinka, T. Gross, W. Unger, Studies of amorphous carbon using X-ray photoelectron spectroscopy, near-edge X-ray-absorption fine structure and Raman spectroscopy, Thin Solid Films. 354 (1999) 106.
[128] M.C. Ferrara, L. Mirenghi, A. Mevoli, L. Tapfer, Synthesis and characterization of sol–gel silica films doped with size-selected gold nanoparticles, Nanotechnology. 19 (2008) 365706.

[129] S.A. Alekseev, V.N. Zaitsev, J. Fraissard, Organosilicas with Covalently Bonded Groups under Thermochemical Treatment, Chemistry of Materials. 18 (2006) 1981.
[130] J.-C. Li, T. Yu, M.-S. Ye, X.-J. Fan, The effects of the vacuum sublimation process on the composition of C_{60}/C_{70} films, Thin Solid Films. 345 (1999) 236.
[131] E.A. Lawton, The Thermal Stability of Copper Phthalocyanine, The Journal of Physical Chemistry. 62 (1958) 384.
[132] K. Wada, K. Tada, N. Itayama, T. Kondo, T.-aki Mitsudo, Preparation of microporous acidic oxides from aluminum-bridged silsesquioxanes and catalytic activities for the cracking of hydrocarbons, Journal of Catalysis. 228 (2004) 374.
[133] H.J. Martin, K.H. Schulz, J.D. Bumgardner, K.B. Walters, XPS Study on the Use of 3-Aminopropyltriethoxysilane to Bond Chitosan to a Titanium Surface, Langmuir. 23 (2007) 6645.
[134] A.K. Chauhan, D.K. Aswal, S.P. Koiry, S.K. Gupta, J.V. Yakhmi, C. Sürgers, et al., Self-assembly of the 3-aminopropyltrimethoxysilane multilayers on Si and hysteretic current–voltage characteristics, Appl. Phys. A. 90 (2007) 581.
[135] S.J. Duclos, R.C. Haddon, S.H. Glarum, A.F. Hebard, K.B. Lyons, The influence of oxygen on the Raman spectrum of C_{60} films, Solid State Commun. 80 (1991) 481.
[136] C.N. Kramer, G. Abrasonis, M.S. Amer, R. Blanco, Z. Chen, Fullerene Research Advances, first ed., Nova Science Publishers, New York, 2007.
[137] S. Bhattacharyya, M. Hietschold, F. Richter, Investigation on the change in structure of tetrahedral amorphous carbon by a large amount of nitrogen incorporation, Diamond Relat. Mater. 9 (2000) 544.
[138] M. Kondo, T.E. Mates, D.A. Fischer, F. Wudl, E.J. Kramer, Bonding structure of phenylacetylene on hydrogen-terminated Si(111) and Si(100): surface photoelectron spectroscopy analysis and ab initio calculations, Langmuir. 26 (2010) 17000.
[139] N. Graf, E. Yegen, T. Gross, A. Lippitz, W. Weigel, S. Krakert, et al., XPS and NEXAFS studies of aliphatic and aromatic amine species on functionalized surfaces, Surf. Sci. 603 (2009) 2849.
[140] A. Schöll, R. Fink, E. Umbach, G.E. Mitchell, S.G. Urquhart, H. Ade, Towards a detailed understanding of the NEXAFS spectra of bulk polyethylene copolymers and related alkanes, Chem. Phys. Lett. 370 (2003) 834.
[141] M. Ramm, M. Ata, K.W. Brzezinka, T. Gross, W. Unger, X-ray photoelectron spectroscopy and near-edge X-ray-absorptionfine structure of C_{60} polymer films, Thin Solid Films. 354 (1999) 106.
[142] M. Plaschke, J. Rothe, M. Altmaier, M.A. Denecke, T. Fanghänel, Near edge X-ray absorption fine structure (NEXAFS) of model compounds for the humic acid/actinide ion interaction, J. Electron Spectrosc. Relat. Phenom. 148 (2005) 151.
[143] T. Bitzer, N.V. Richardson, S. Reiss, M. Wühn, C. Wöll, Sodium-induced ordering of the benzoate species on Si(100)-2×1: a combined HREELS, XPS and NEXAFS study, Surf. Sci. 458 (2000) 173.
[144] S.C. Ray, C.W. Pao, J.W. Chiou, H.M. Tsai, J.C. Jan, W.F. Pong, et al., Electronic properties of a-CN_x thin films: An x-ray-absorption and photoemission spectroscopy study, J. Appl. Phys. 98 (2005) 033708.
[145] A.J. Maxwell, P.A. Brühwiler, D. Arvanitis, J. Hasselström, M.K.J. Johansson, N. Mårtensson, Electronic and geometric structure of C_{60} on Al(111) and Al(110), Phys. Rev. B. 57 (1998) 7312.
[146] R.R. Cooney, S.G. Urquhart, Chemical trends in the near-edge x-ray absorption fine structure of monosubstituted and para-bisubstituted benzenes, J. Phys. Chem. B. 108 (2004) 18185.
[147] S. Diegoli, P.M. Mendes, E.R. Baguley, S.J. Leigh, P. Iqbal, Y.R. Garcia Diaz, et al., pH-Dependent gold nanoparticle self-organization on functionalized Si/SiO_2 surfaces, J. Exp. Nanosci. 1 (2006) 333.
[148] T. Maeda, A. Kurokawa, K. Sakamoto, A. Ando, H. Itoh, S. Ichimura, Atomic force microscopy observation of layer-by-layer growth of ultrathin silicon dioxide by ozone gas at room

temperature, J. Vac. Sci. Technol. B. 19 (2001) 589.
[149] E.L. Hanson, J. Schwartz, B. Nickel, N. Koch, M.F. Danisman, Bonding Self-Assembled, Compact Organophosphonate Monolayers to the Native Oxide Surface of Silicon, J. Am. Chem. Soc. 125 (2003) 16074.
[150] D.J. Whitehouse, D.J. Whitehouse, Handbook of Surface and Nanometrology, first ed., Institute of Physics Publishing, 2003.
[151] E.S. Gadelmawla, M.M. Koura, T.M.A. Maksoud, I.M. Elewa, H.H. Soliman, Roughness parameters, J. Mater. Process. Technol. 123 (2002) 133.
[152] T. Komeda, K. Namba, Y. Nishioka, Self-assembled-monolayer film islands as a self-patterned-mask for SiO_2 thickness measurement with atomic force microscopy, Appl. Phys. Lett. 70 (1997) 3398.
[153] Y. Cai, B.-min Z. Newby, Dewetting of Polystyrene Thin Films on Poly(ethylene glycol)-Modified Surfaces as a Simple Approach for Patterning Proteins, Langmuir. 24 (2008) 5202.
[154] X. Yu, J. Zhang, W. Choi, J.-Y. Choi, J.M. Kim, L. Gan, et al., Cap Formation Engineering: From Opened C_{60} to Single-Walled Carbon Nanotubes, Nano Lett. 10 (2010) 3343.
[155] S.-ichiro Kobayashi, Y. Cho, Investigation of interface between fullerene molecule and Si(111)-7×7 surface by noncontact scanning nonlinear dielectric microscopy, J. Vac. Sci. Technol. B. 28 (2010) C4D18.
[156] J.A. Heredia-Guerrero, M.A. San-Miguel, M.S.P. Sansom, A. Heredia, J.J. Benítez*, Chemical Reactions in 2D: Self-Assembly and Self-Esterification of 9(10),16-Dihydroxypalmitic Acid on Mica Surface, Langmuir. 25 (2009) 6869.
[157] J.D. Cogdell, A convolved multi-Gaussian probability distribution for surface topography applications, Precis. Eng. 32 (2008) 34.
[158] I. Salzmann, S. Duhm, R. Opitz, R.L. Johnson, J.P. Rabe, N. Koch, Structural and electronic properties of pentacene-fullerene heterojunctions, J. Appl. Phys. 104 (2008) 114518.
[159] S. Yoo, W.J. Potscavage Jr., B. Domercq, S.-H. Han, T.-D. Li, S.C. Jones, et al., Analysis of improved photovoltaic properties of pentacene/C_{60} organic solar cells: Effects of exciton blocking layer thickness and thermal annealing, Solid-State Electron. 51 (2007) 1367.
[160] K. Kohli, H. Chaudhary, P. Rathee, S. Rathee, V. Kumar, Fullerenes: New Contour to Carbon Chemistry, Pharma Times 41 (2009) 9.
[161] K. Broczkowska, J. Krzak-Roś, M. Miller, in: Preliminary Study of Fullerene Doped Thin Films Obtained by Sol-gel Method, 2009: pp. 61.
[162] H. Ningkang, Y. Bing, W. Dezhi, Electron spectroscopy studies on SiC films before and after hydrogen ion irradiation, Journal of Wuhan University of Technology, Materials Science Edition. 20 (2005) 1.
[163] T. Uno, H. Tabata, T. Kawai, Peptide- Nucleic Acid-Modified Ion-Sensitive Field-Effect Transistor-Based Biosensor for Direct Detection of DNA Hybridization, Anal. Chem. 79 (2007) 52.
[164] F. Parmigiani, L. Depero, Diffraction and XPS studies of Cu complexes of intercalated compounds of α-zirconium phosphate. II: XPS electronic structures, Structural Chemistry. 5 (1994) 117.
[165] A. Naumkin, A. Krasnov, E. Said-Galiev, I. Volkov, A. Nikolaev, O. Afonicheva, et al., Carbon dioxide in the surface layers of ultrahigh molecular weight polyethylene, Doklady Physical Chemistry. 419 (2008) 68.
[166] M. Krzywiecki, L. Grządziel, L. Ottaviano, P. Parisse, S. Santucci, J. Szuber, XPS study of air exposed copper phthalocyanine ultra-thin films deposited on Si(111) native substrates, Materials Science-Poland. 26 (2008) 287.
[167] W. Chen, Q. Han, R. Most, C. Waldfried, O. Escorcia, I. Berry, Plasma Impacts to an O-SiC Low-k Barrier Film, J. Electrochem. Soc. 151 (2004) F182.
[168] Y.-S. Li, Y. Wang, S. Ceesay, Vibrational spectra of phenyltriethoxysilane, phenyltrimethoxysilane and their sol-gels, Spectrochimica Acta Part A: Molecular and Biomolecular Spectroscopy. 71 (2009) 1819.

[169] K. Kolanek, M. Tallarida, K. Karavaev, D. Schmeisser, In situ measurements of the atomic layer deposition of high-k dielectrics by atomic force microscope for advanced microsystems, in: Students and Young Scientists Workshop "Photonics and Microsystems", 2009 International, 2009: pp. 47.
[170] N.G. Orji, R.G. Dixson, J. Fu, T.V. Vorburger, Traceable pico-meter level step height metrology, Wear. 257 (2004) 1264.
[171] P. Hoffmann, D. Schmeisser, H.-J. Engelmann, E. Zschech, H. Stegmann, F. Himpsel, et al., Characterization of Chemical Bonding in Low-k Dielectric Materials for Interconnect Isolation: A XAS and EELS Study, in: Mater. Res. Soc. Symp. Proc., 2006: pp. F01.
[172] K. Zagorodniy, D. Chumakov, C. Taschner, A. Lukowiak, H. Stegmann, D. Schmeisser, et al., Novel Carbon-Cage-Based Ultralow-k Materials: Modeling and First Experiments, IEEE Trans. Semicond. Manufact. 21 (2008) 646.
[173] N. Satyanarayana, S.K. Sinha, L. Shen, Effect of Molecular Structure on Friction and Wear of Polymer Thin Films Deposited on Si Surface, Tribol. Lett. 28 (2007) 71.
[174] D.G. Kurth, T. Bein, Thin Films of (3-Aminopropyl)triethoxysilane on Aluminum Oxide and Gold Substrates, Langmuir. 11 (1995) 3061.
[175] B. Todorovic Markovic, V. Jokanovic, S. Jovanovic, D. Kleut, M. Dramicanin, Z. Markovic, Surface chemical modification of fullerene by mechanochemical treatment, Appl. Surf. Sci. 255 (2009) 7537.
[176] A.J. Nolte, J.Y. Chung, M.L. Walker, C.M. Stafford, In situ Adhesion Measurements Utilizing Layer-by-Layer Functionalized Surfaces, ACS Appl. Mater. Interfaces. 1 (2009) 373.
[177] X. Wang, Y.H. Tseng, J.C.C. Chan, S. Cheng, Direct synthesis of highly ordered large-pore functionalized mesoporous SBA-15 silica with methylaminopropyl groups and its catalytic reactivity in flavanone synthesis, Microporous Mesoporous Mater. 85 (2005) 241.
[178] E. Halpern, B. Khamaisi, O. Shaya, G. Shalev, I. Levy, Y. Rosenwaks, Electrostatic Properties of Silane Monolayers in an Electrolytic Environment, J. Phys. Chem. C. 113 (2009) 16802.
[179] K. Broczkowska, J. Klocek, D. Friedrich, K. Henkel, K. Kolanek, A. Urbanowicz, et al., Fullerene based materials for ultra-low-k application, in: Students and Young Scientists Workshop "Photonics and Microsystems", 2010 International, 2010: pp. 39.
[180] D. Schmeisser, K. Henkel, K. Müller, M. Tallarida, Interface Reactions in Ultrathin Functional Dielectric Films, Adv. Eng. Mater. 11 (2009) 269.
[181] K.S. Chen, S.C. Chen, Y.C. Yeh, W.C. Lien, H.R. Lin, J.M. Yang, The Study of Immobilization Thermal-Sensitive Hydrogel onto ePTFE Film Use the Cold Plasma and Photo-Grafting Technique, Advanced Materials Research. 15 (2007) 187.
[182] A. Bendeddouche, R. Berjoan, E. Beche, T. Merle-Mejean, S. Schamm, V. Serin, et al., Structural characterization of amorphous SiCxNy chemical vapordeposited coatings, J. Appl. Phys. 81 (2009) 6147.
[183] K. Akaike, K. Kanai, Y. Ouchi, K. Seki, Influence of side chain of [6,6]-phenyl-C[sub 61]-butyric acid methyl ester on interfacial electronic structure of [6,6]-phenyl-C_{61}-butyric acid methyl ester /Ag substrate, Appl. Phys. Lett. 94 (2009) 043309.
[184] J. Fu, V. Tsai, R. Köning, R. Dixson, T. Vorburger, Algorithms for calculating single-atom step heights, Nanotechnology. 10 (1999) 428.
[185] H. Edwards, R. McGlothlin, E. U, Vertical metrology using scanning-probe microscopes: Imaging distortions and measurement repeatability, J. Appl. Phys. 83 (1998) 3952.
[186] L.I. Fedina, D.V. Sheglov, S.S. Kosolobov, A.K. Gutakovskii, A.V. Latyshev, Precise surface measurements at the nanoscale, Meas. Sci. Technol. 21 (2010) 054004.
[187] M. Suzuki, S. Aoyama, T. Futatsuki, A.J. Kelly, T. Osada, A. Nakano, et al., Standardized procedure for calibrating height scales in atomic force microscopy on the order of 1 nm, J. Vac. Sci. Technol. A. 14 (2009) 1228.
[188] M. Mehlhorn, K. Morgenstern, Height analysis of amorphous and crystalline ice structures on Cu (111) in scanning tunneling microscopy, New J. Phys. 11 (2009) 093015.

[189] M. Uematsu, E.U. Franck, Static Dielectric Constant of Water and Steam, J. Phys. Chem. Ref. Data. 9 (1980) 1291.
[190] J. Jordan, K.I. Jacob, R. Tannenbaum, M.A. Sharaf, I. Jasiuk, Experimental trends in polymer nanocomposites-a review, Mater. Sci. Eng., A. 393 (2005) 1.
[191] L. Merhari, Hybrid Nanocomposites for Nanotechnology: Electronic, Optical, Magnetic and Biomedical Applications, first ed., Springer, 2009.
[192] A.S. Abd-El-Aziz, C.E. Carraher, C.U. Pittman, M. Zeldin, Macromolecules Containing Metal and Metal-Like Elements, Group IVA Polymers: Volume 4, first ed., Wiley-Interscience, 2005.
[193] K. Liang, G. Li, H. Toghiani, J.H. Koo, C.U. Pittman, Cyanate Ester/Polyhedral Oligomeric Silsesquioxane (POSS) Nanocomposites: Synthesis and Characterization, Chem. Mater. 18 (2006) 301.
[194] R. Verker, E. Grossman, I. Gouzman, N. Eliaz, TriSilanolPhenyl POSS-polyimide nanocomposites: Structure-properties relationship, Compos. Sci. Technol. 69 (2009) 2178.
[195] G. Li, L. Wang, H. Ni, C.U. Pittman, Polyhedral oligomeric silsesquioxane (POSS) polymers and copolymers: a review, J. Inorg. Organomet. Polym. 11 (2001) 123.
[196] L. Zhang, H.C.L. Abbenhuis, Q. Yang, Y.-M. Wang, P.C.M.M. Magusin, B. Mezari, et al., Mesoporous Organic–Inorganic Hybrid Materials Built Using Polyhedral Oligomeric Silsesquioxane Blocks, Angew. Chem. Int. Ed. 46 (2007) 5003.
[197] K. Yamamoto, H. Itoh, XPS study of silicon surface after ultra-low-energy ion implantation, Surf. Sci. 600 (2006) 3753.
[198] I. George, P. Viel, C. Bureau, J. Suski, G. Lécayon, Study of the Silicon/γ-APS/Pyralin Assembly Interfaces by X-ray Photoelectron Spectroscopy, Surf. Interface Anal. 24 (1996) 774.
[199] D. Joung, A. Chunder, L. Zhai, S.I. Khondaker, High yield fabrication of chemically reduced graphene oxide field effect transistors by dielectrophoresis, Nanotechnology. 21 (2010) 165202.
[200] C.K.M. Heo, J.W. Bunting, Nucleophilicity towards a vinylic carbon atom: rate constants for the addition of amines to the 1-methyl-4-vinylpyridinium cation in aqueous solution, J. Chem. Soc., Perkin Trans. 2. (1994) 2279.
[201] D.A. Buttry, J.C.M. Peng, J.-B. Donnet, S. Rebouillat, Immobilization of amines at carbon fiber surfaces, Carbon. 37 (1999) 1929.
[202] J. Klocek, K. Henkel, K. Kolanek, K. Broczkowska, D. Schmeißer, M. Miller, et al., Studies of chemical and electrical properties of fullerene and 3-aminopropyl-trimethoxysilane based low-k materials, Thin Solid Films. (submitted).
[203] M. Pelliccione, T.-M. Lu, Evolution of Thin Film Morphology: Modeling and Simulations, first ed., Springer, 2007.
[204] K. Liang, G. Li, H. Toghiani, J.H. Koo, C.U. Pittman, Cyanate Ester/Polyhedral Oligomeric Silsesquioxane (POSS) Nanocomposites: Synthesis and Characterization, Chem. Mater. 18 (2006) 301.
[205] P. Meakin, Fractals, Scaling and Growth Far from Equilibrium, first ed., Cambridge University Press, 1998.
[206] P. Meakin, Diffusion-controlled aggregation on two-dimensional square lattices: Results from a new cluster-cluster aggregation model, Phys. Rev. B. 29 (1984) 2930.
[207] F. Family, P. Meakin, T. Vicsek, Cluster size distribution in chemically controlled cluster–cluster aggregation, J. Chem. Phys. 83 (1985) 4144.
[208] P. Meakin, The growth of rough surfaces and interfaces, Phys. Rep. 235 (1993) 189.
[209] S. Hüfner, Photoelectronspectroscopy, third ed., Springer, Berlin, Germany, 2003.
[210] J.A. Howarter, J.P. Youngblood, Optimization of Silica Silanization by 3-Aminopropyltriethoxysilane, Langmuir. 22 (2006) 11142.
[211] M. Chelvayohan, C.H.B. Mee, Work function measurements on (110), (100) and (111) surfaces of silver, J. Phys. C: Solid State Phys. 15 (1982) 2305.

[212] C.-W. Kim, J.C. Villagrán, U. Even, J.C. Thompson, Adsorbate effects on photoemission from Ag, J. Chem. Phys. 94 (1991) 3974.
[213] W. Volksen, R.D. Miller, G. Dubois, Low Dielectric Constant Materials, Chem. Rev. 110 (2010) 56.
[214] D. Hiller, R. Zierold, J. Bachmann, M. Alexe, Y. Yang, J.W. Gerlach, et al., Low temperature silicon dioxide by thermal atomic layer deposition: Investigation of material properties, J. Appl. Phys. 107 (2010) 064314.
[215] J. Klocek, K. Kolanek, K. Henkel, E. Zschech, D. Schmeisser, Influence of the fullerene derivatives and cage polyhedral oligomeric silsesqiuoxanes on 3-aminopropyltrimethoxysilane based hybrid nanocomposites chemical, morphological and electrical properties, Surface Science. (submitted).
[216] N.Y. Borovkov, S.V. Blokhina, L.A. Val'kova, M.V. Ol'khovich, G.V. Sibrina, A.V. Sharapova, Interactions of copper tetra-tert-butylphthalocyanine with nitrogen-and sulfur-containing compounds in solutions, Russian Chemical Bulletin. 52 (2003) 1522.

Appendix

Frequently used abbreviations and symbols

A	area	I	intensity
AFM	atomic force microscopy	k	relative dielectric constant
APTMS	3-aminopropyl-trimethoxysilane	*LUMO*	lowest unoccupied molecular orbital
CuPc	copper phthalocyanine		
C	capacitance	*MIS*	metal-insulator-semiconductor
C_{acc}	accumulation capacitance	*NEXAFS*	near edge X-ray absorption fine structure spectroscopy
CV	capacitance voltage		
d	thickness	*PCBM*	[6,6]-phenyl-C_{61}-butyric acid
E_B	binding energy	*POSS*	tris(dimethylvinylsilyloxy)-POSS
E_F	Fermi energy	*RMS*	root mean square
E_k	kinetic energy	*TEY*	total electron yield
E_V	valence band energy (maximum)	*TFY*	total fluorescence yield
ε_0	vacuum permittivity	S_{ku}	surface kurtosis
FTIR	Fourier Transform Infrared Spectroscopy	S_q	RMS surface roughness
		S_{sk}	surface skewness
H_2O_2	hydrogen peroxide	v/v	volume per volume
HOMO	highest occupied molecular orbital	*XPS*	X-ray photoelectron spectroscopy
H_2SO_4	sulfuric acid	ϕ	work function
$h\nu$	photon energy	λ	mean free path

Acknowledgements

First of all I would like to thank my advisor, Prof. Dieter Schmeißer, without whom this thesis would not exist now since he is the person who gave me the great opportunity to realize all researches at BTU Cottbus. At the same time he permitted me to preserve the necessary degree of freedom in the performed studies not limiting my creativity but supporting it by the fruitful discussions.

I am very grateful to Prof. Zschech not only for his readiness to become a referee of this thesis and for the fruitful discussion but first of all for the very interesting phenomenon related to his extraordinary personality, which is motivating me to work effectively every time I see him. I would like to thank my second referee Prof. Mirosław Miller for a nice and fruitful collaboration regarding the fullerene containing sol-gel films.

It is not possible to express in a few words my gratitude for two irreplaceable and reliable mentors of this thesis: Dr. Karsten Henkel and Dr. Krzysztof Kolanek. Besides their extraordinary ability to work in a group they are also amazingly skilled experts in their fields (Karsten-CV, Krzysztof-AFM). There is no other possibility: collaboration with them must be successful.

Special thanks to Guido Beuckert for a professional and efficient technical support and obviously for helping me to gather the first experience in the field of XPS.

I truly appreciate the contribution of Daniel Friedrich, who bravely survived long nights at the synchrotron while recording the NEXAFS spectra.

Greatly acknowledgment is the contribution of the group from Wrocław University of Technology created by: Prof. Miroslaw Miller, Dr. Justyna Krzak-Roś, Katarzyna Broczkowska, Dr. Adam Urbanowicz. The sol-gel film synthesized by them and their experience accelerated significantly the progress of the experiments described in this thesis.

I would like to thank Matthias Richter, Dr. Massimo Tallarida, Dr. Klaus Müller, Shine Philip, Ioanna Paloumpa, Carola Schwiertz, David Hoffmannbeck, Marcel Michling, Matthias Städter, Roy Mbua, Matthias Kraatz, Sebastian Müller, Konstantin Karavaev, Jakub Wyrodek, Łukasz Starzyk, Krzysztof Kachel, Marcin Wróblewski and every colleagues I'd worked with at BTU for creating kind and nice working atmosphere. It was a real pleasure to work with them. Especially I would like to thank Mattias Richter, not only for the "skillful assistance during the contact evaporation" but also for his precious time, that he never hesitate to reserve in order to share his great scientific knowledge with other, less experienced colleagues. Special thanks to Shine for taking care of the plants in LB 1B and Ioanna for the creation of the sympathetic climate of this place.

Finally I would like to thank Karla Kersten for being the best secretary I've ever met and Susanne Masch for helping with the administrative problems.

i want morebooks!

Buy your books fast and straightforward online - at one of world's fastest growing online book stores! Environmentally sound due to Print-on-Demand technologies.

Buy your books online at
www.get-morebooks.com

Kaufen Sie Ihre Bücher schnell und unkompliziert online – auf einer der am schnellsten wachsenden Buchhandelsplattformen weltweit! Dank Print-On-Demand umwelt- und ressourcenschonend produziert.

Bücher schneller online kaufen
www.morebooks.de

VDM Verlagsservicegesellschaft mbH
Heinrich-Böcking-Str. 6-8
D - 66121 Saarbrücken

Telefon: +49 681 3720 174
Telefax: +49 681 3720 1749

info@vdm-vsg.de
www.vdm-vsg.de

Printed by Books on Demand GmbH, Norderstedt / Germany